Contents

Heating,
Ventilating
and
Air Conditioning

Heating,
Ventilating
and
Air Conditioning

F. Hall, MIOB, MIPHE

The Construction Press
Lancaster, London and New York

The Construction Press Ltd, Lunesdale House, Hornby, Lancaster, England

A subsidiary company of Longman Group Ltd, London.
Associated companies, branches and representatives throughout the world.

Published in the United States of America by Longman Inc, New York.

© F Hall, 1980

First published 1980

British Library Cataloguing in Publication Data

Hall, Fred
    Heating, ventilating and air conditioning.
    1. Heating – Equipment and supplies
    2. Air conditioning – Equipment and supplies
    3. Ventilation – Equipment and supplies
    I. Title
    697   TH7011
    ISBN 0-86095-884-1

Printed in Great Britain by Fletcher and Son Limited, Norwich

Publishers' Note

The contents of this book were originally published in the author's larger three-volume work entitled Building Services and Equipment (1976 and 1979). These chapters on heating, ventilating and air conditioning have been produced in this single volume reference book to meet the well established need for such a volume.

Acknowledgements

We are grateful to the following for permission to reproduce copyright material published in the original three volumes: British Standards Institution; Building Research Establishment; The Chartered Institution of Building Services; The Controller of Her Majesty's Stationery Office; The Illuminating Engineering Society; The Institute of Plumbing; The Institution of Heating and Ventilating Engineers; Lighting Industry Federation Ltd and The Wednesbury Tube Company.

We are also grateful to the following for permission to reproduce diagrams: British Gas Corporation; Building Research Establishment; The Walter Kidde Co Ltd; Klargester Environmental Engineering Limited; Marley Plumbing; Marscar Limited; Mather and Platt Ltd; Spirax Limited; Wavin Plastics Limited.

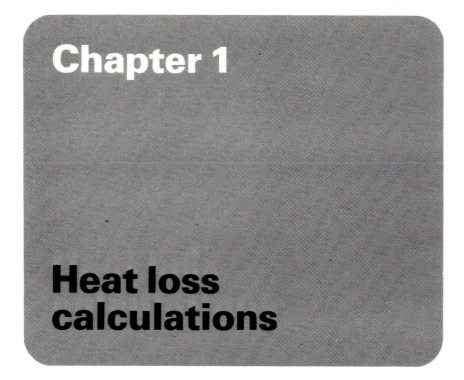

Chapter 1

Heat loss calculations

Heat loss calculations

Definition of terms

1. Thermal conductivity (k)
The thermal transmission in unit time through unit area of a slab, or a uniform homogeneous material of unit thickness, when the difference of temperature is established between its surfaces.

The unit is W/m °C

2. Thermal resistivity (r)
The reciprocal of the thermal conductivity.

The unit is m °C/W

3. Thermal conductance (c) in
The thermal transmission is unit time, through a unit area of a uniform structural component of thickness L per unit of temperature difference, between the hot and cold surfaces.

The unit is W/m² °C

4. Thermal resistance (R)
The reciprocal of the thermal conductance.

$$R = \frac{L}{k}$$

where:

R = Thermal resistance of a material (m² °C/W)

L = Thickness of the material in metres

k = Thermal conductivity of the material (W/m °C)

Note: If the thickness L is given in millimetres, it must be converted to metres.

5. Thermal transmission (U)
The thermal transmission in unit time through unit area of a given structure, divided by the difference between the environmental temperature on either side of the structure.

The unit is W/m² °C

6. Standard thermal transmission (Standard U)
The value of the thermal transmission of a building element related to standard conditions.

7. Design thermal transmission (Design U)
The value of the thermal transmission of a building element for prevailing design conditions.

8. Emissivity (e)
The ratio of the thermal radiation from unit area of a surface, to the radiation from unit area of a black body at the same temperature.

9. Environmental temperature (t_e)
A balanced mean, between the mean radiant temperature and the air temperature. It may be evaluated approximately from the following formula:

$$t_{ei} = \frac{2}{3} t_{ri} + \frac{1}{3} t_{ai}$$

where

t_{ri} = mean radiant temperature of all the room surfaces in °C

t_{ai} = inside air temperature in °C

The concept of environmental temperature for heat loss calculations provides a more accurate assessment of steady state heat loss than the conventional procedure that uses air temperature as a basis. The internal environmental temperature is also better than the internal air temperature as an index of the thermal comfort of the internal environment. This places the designer in a more favourable position for assessing the internal thermal comfort than is possible employing the air temperature. Conventional methods however may be followed when the difference between the mean radiant temperature and the air temperature is quite small. This occurs when rooms have little exposure to the outside and the standard of building thermal insulation is very high.

10. Sol-air temperature (t_{eo})
The outside air temperature which in the absence of solar radiation would give the same temperature distribution and rate of heat transfer, through the wall or

roof, as exists with the actual outdoor temperature and the incidence of solar radiation.

11. *Mean radiant temperature* (t_{rm})

The temperature of a uniform block enclosure, in which a solid body or occupant would exchange the same amount of radiant heat as in existing non-uniform environment.

12. *Degrees Kelvin* (K)

This is defined in terms of the triple point of water as the fundamental fixed point, attributing to it the temperature of 273.16 K. Absolute zero is defined as 0 K.

13. *Degrees Celsius* (°C)

Degrees Celsius 0 °C = 273.16 K and the intervals on the Celsius and Kelvin scales are identical.

14. *Heat*

Heat is a form of energy, its quantity is measured in joules (J).

15. *Power*

Power is measured in watts (W). One watt is equal to one joule per second.

16. *Heat transfer*

There are three ways in which heat may be transferred from a material: (*a*) conduction, (*b*) convection and (*c*) radiation.

Conduction. The molecules of a material at a higher temperature will vibrate more than the molecules of a material at a lower temperature and this vibrational heat energy is transferred from the higher temperature to the lower temperature. This heat transfer takes place without movement of the hotter molecules to the cooler molecules, and the greater the temperature difference the greater the transfer of heat by conduction.

Conduction is greater in solids than in gases. Still air conducts heat very slowly and an unventilated cavity provides a good insulator. Still air pockets in insulating materials provide good heat-insulating characteristics.

Convection: this is the transfer of heat in a fluid. The hotter less dense liquid or gas is displaced by the more dense liquid or gas surrounding it, thus creating circulation. The natural circulation of air in a room, or water in a heating system, is brought about by convection.

Radiation: this is the transfer of heat in the form of electro-magnetic radiation from one body to another, without the need of a conducting medium. All bodies emit radiant heat and receive radiant heat back from other bodies. The higher the temperature of the body, the greater the radiant heat emitted. Matt black surfaces generally radiate or receive more heat than white, or bright shiny surfaces. White chippings on roofs will help to prevent a roof slab receiving radiant heat from the sun and aluminium foil will reflect radiant heat back from a heated room and this will act as a heat insulator. Figure 1.1 shows the transfer of heat through a wall by conduction, convection and radiation.

Temperature distribution

Figure 1.2 shows the temperature distribution for a solid wall. The wall with a low thermal conductivity will have a higher outside surface temperature than the wall with a high thermal conductivity and therefore the temperature distribution will be greater. Figure 1.3 shows the temperature distribution for a cavity wall with a plastered aerated inner leaf, with a low thermal conductivity and a brick external leaf, and a fairly high thermal conductivity. In Fig. 1.4 the cavity is filled with urea formaldehyde foam, which will improve the thermal insulation and the temperature distribution through the cavity will be greater than the air cavity.

Standard *U* values

Steady state conditions

The rate of transfer of heat through a material may be affected by:

(*a*) Amount of moisture in the material.
(*b*) The variations in the composition of the material.
(*c*) Jointing of the component parts.

In heat loss calculations, however, it is assumed that a steady state exists which would result if the material is homogeneous; for example, it has the same composition throughout and the material is also dry, due to receiving heat from inside the building.

Computation of *U* values

The thermal transmission through the structure is obtained by combining the thermal resistance of its components and the adjacent air layers. The thermal transmission is found by adding the thermal resistances and taking the reciprocal.

$$U = \frac{1}{R_{si} + R_{so} + R_1 + R_2 + R_3 + R_a}$$

where

U = thermal transmission W/m² °C

R_{si} = inside surface resistance m² °C/W

R_{so} = outside surface resistance m² °C/W

R_1 = thermal resistance of the structural components m² °C/W

R_2 = thermal resistance of the structural components m² °C/W

R_3 = thermal resistance of the structural components m² °C/W

R_a = resistance of air space m² °C/W

In computation of *U* values, the thermal resistances L/k are used. Where

k = thermal conductivity W/m °C

L = thickness in (m) of a uniform homogeneous material

$$U = \frac{1}{R_{si} + R_{so} + \dfrac{L_1}{k_1} + \dfrac{L_2}{k_2} + \dfrac{L_3}{k_3} + R_a}$$

Table 1.1 gives the thermal conductivity for commonly used building materials.

Table 1.1 Thermal conductivities of common building materials

Material	k value W/m °C
Asbestos cement sheet	0.40
Asbestos insulating board	0.12
Asphalt	1.20
Brickwork (commons)	
light	0.80
average	1.20
dense	1.47
Brickwork (engineering)	1.15
Concrete	
structural	1.40
aerated	0.14
Cork slab	0.40
Clinker block	0.05
Glass	1.02
Glass wool	0.034
Gypsum plasterboard	0.15—0.58
Linoleum (inlaid)	0.20
Plastering	
gypsum	0.40
vermiculite	0.20
Plywood	0.138
Polystyrene foam slab	0.04
Polyurethane foam	0.020—0.025
Foamed urea formaldehyde	0.030—0.036
Hair felt	0.43
Rendering (cement and sand)	0.53
Roofing felt	0.20
Rubber flooring	0.40
Slates	1.50
Soil	1.00—1.15
Stone	
granite	2.90
limestone	1.50
sandstone	1.30
Strawboard	0.09
Tiles	
burnt clay	0.83
plastic	0.50
Timber	
softwood	0.14
hardwood	0.16
Vermiculite	0.36—0.58
Wood	
chipboard	0.108
wool slabs	0.09

Surface resistances

The transfer of heat by convection to or from a homogeneous material depends upon the velocity at which the air passes over the surface of the material and the roughness of the surface. An internal wall may have a smooth plastered surface with very little air movement and an external wall may have a rough exterior surface and a high air movement.

The smooth internal surface will have a small amount of heat transfer by convection. The air forms a stagnant film, which tends to insulate the wall surface from the warmer air in the room. On the external surface of the material, the wind forces acting on a rough surface will cause eddy currents and heat will be transferred at a higher rate. Table 1.2 gives the inside surface resistances.

Table 1.2 Inside surface resistances R_{si} in m^2 °C/W ·

Building element	Heat flow	Surface resistance in m^2 °C/W	
		High emissivity	Low emissivity
Walls	Horizontal	0.123	0.304
Ceilings, flat or pitched roofs, floors	Upward	0.106	0.218
Ceilings and floors	Downward	0.150	0.562

Table 1.3 gives the external surface resistance.

Table 1.3 External surface resistance R_{so} in m^2 °C/W for various exposures and surfaces

Building element	Emissivity of surface	Surface resistance for stated exposure (m^2 °C/W)		
		Sheltered	Normal (standard)	Severe
Wall	High	0.08	0.055	0.03
	Low	0.11	0.067	0.03
Roof	High	0.07	0.045	0.02
	Low	0.09	0.053	0.02

Sheltered: up to third-floor buildings in city centres.
Normal: most suburban and country premises and the fourth to eighth floors, of buildings in city centres.
Severe: buildings on the coast, or exposed hill sites and above the fifth floor of buildings in suburban, or country districts, or above the ninth floor of buildings in city centres.

Thermal resistances of air spaces

Air spaces can be regarded as a further thermal resistance and an unventilated space offers more resistance than a ventilated one. Table 1.4 gives the standard unventilated resistances of unventilated air spaces.

Table 1.4

Air space thickness	Surface emissivity	Thermal resistance (m^2 °C/W)	
		Heat flow horizontal or upwards	Heat flow downwards
5 mm	High	0.11	0.11
	Low	0.18	0.18
20 mm	High	0.18	0.21
	Low	0.35	1.06

Example 1.1 (Fig. 1.5). *A flat roof consists of 150 mm thick concrete covered by 20 mm of asphalt and sheltered conditions exists. Calculate the thermal transmittance (U) for the roof.*

Thermal conductivities
Concrete   1.4 W/m °C
Asphalt   1.2 W/m °C

Thermal resistances
Inside surface   0.11 m^2 °C/W
Outside surface   0.07 m^2 °C/W

$$U = \frac{1}{R_{si} + R_{so} + \dfrac{L_1}{k_1} + \dfrac{L_2}{k_2}}$$

$$U = \frac{1}{0.11 + 0.07 + \dfrac{0.15}{1.4} + \dfrac{0.02}{1.2}}$$

$$U = \frac{1}{0.11 + 0.07 + 0.11 + 0.02}$$

$$U = \frac{1}{0.31}$$

$$U = \underline{3.23 \text{ W/m}^2 \text{ °C}}$$

Example 1.2 (Fig. 1.6). *A cavity wall consists of 13 mm thick plaster, 80 mm thick aerated concrete, 50 mm wide unventilated air space and 115 mm of external facing brick. Calculate the thermal transmission (U) for the wall.*

Thermal conductivities
Plaster   0.4 W/m °C
Cellular concrete 0.25 W/m °C
Brick   1.15 W/m °C

Thermal resistances
Inside surface   0.123 m^2 °C/W
Air space   0.18 m^2 °C/W
Outside surface   0.055 m^2 °C/W

$$U = \frac{1}{R_{si} + R_{so} + \dfrac{L_1}{k_1} + \dfrac{L_2}{k_2} + \dfrac{L_3}{k_3} + R_a}$$

$$U = \frac{1}{0.123 + 0.055 + \dfrac{0.013}{0.4} + \dfrac{0.08}{0.25} + \dfrac{0.115}{1.15} + 0.18}$$

$$U = \frac{1}{0.123 + 0.055 + 0.0325 + 0.32 + 0.1 + 0.18}$$

$$U = \frac{1}{0.8105}$$

$$U = \underline{1.234 \text{ W/m}^2 \text{ °C}}$$

Example 1.3 (Fig. 1.7). *One wall of a building consists of 25 mm thickness of cedar boarding, 76 mm thickness of glass wool insulation and 13 mm thickness of plasterboard. Compare the overall thermal transmission of the wall, with the wall given in Example 1.2 .*

Thermal conductivity
Cedar wood   0.14 W/m °C
Glass wool   0.042 W/m °C
Plasterboard   0.58 W/m °C

Thermal resistances
Inside surface   0.123 m^2 °C/W
Outside surface   0.055 m^2 °C/W

$$U = \frac{1}{R_{si} + R_{so} + \dfrac{L_1}{k_1} + \dfrac{L_2}{k_2} + \dfrac{L_3}{k_3}}$$

$$U = \frac{1}{0.123 + 0.055 + \dfrac{0.013}{0.58} + \dfrac{0.076}{0.042} + \dfrac{0.025}{0.14}}$$

$$U = \frac{1}{0.123 + 0.055 + 0.0224 + 1.809 + 0.178}$$

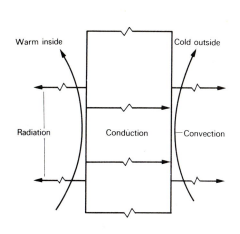

Fig. 1.1 Heat transfer through a wall

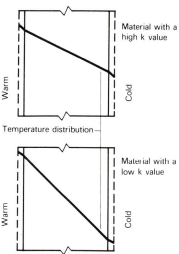

Fig. 1.2 Temperature distribution through wall

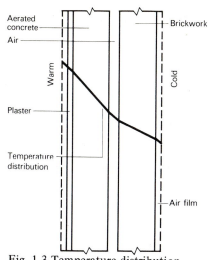

Fig. 1.3 Temperature distribution with air cavity

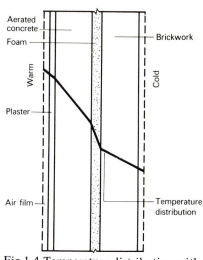

Fig 1.4 Temperature distribution with urea formaldehyde foam filled cavity

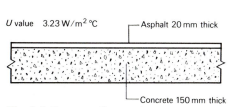

U value   3.23 W/m² °C

Asphalt 20 mm thick

Concrete 150 mm thick

Fig. 1.5 Concrete flat roof covereds covered with ashpalt

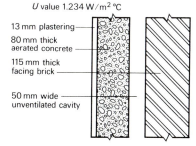

U value 1.234 W/m² °C

13 mm plastering
80 mm thick aerated concrete
115 mm thick facing brick
50 mm wide unventilated cavity

Fig 1.6 Cavity wall — aerated concrete — brickwork

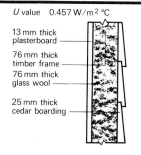

U value   0.457 W/m² °C

13 mm thick plasterboard
76 mm thick timber frame
76 mm thick glass wool
25 mm thick cedar boarding

Fig 1.7 Lightweight wall

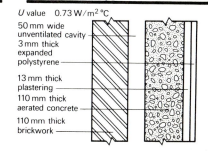

U value   0.73 W/m² °C

50 mm wide unventilated cavity
3 mm thick expanded polystyrene
13 mm thick plastering
110 mm thick aerated concrete
110 mm thick brickwork

Fig. 1.8 Cavity wall — aerated concrete — brickwork and expanded polystyrene

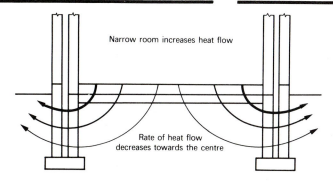

Narrow room increases heat flow

Rate of heat flow decreases towards the centre

Fig. 1.9

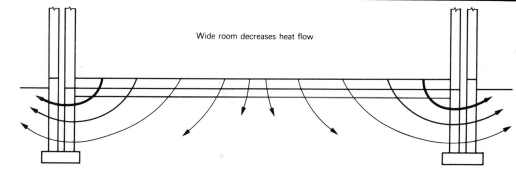

Wide room decreases heat flow

Fig. 1.10 Heat flow through solid ground floor

$$U = \frac{1}{2.188}$$

$U = 0.457$

$U = 0.457 \text{ W/m}^2 \,^\circ\text{C}$

If the inside and outside environmental temperatures are 20 °C and −2 °C respectively, the rate of heat loss per metre squared would be:

For solid wall

Heat loss = 1.234 x 1.0 x [20 − (−2)]

Heat loss = 27.148 W

For lightweight wall

Heat loss = 0.457 x 1.0 x [20 − (−2)]

Heat loss = 10.054 W

The heat loss through the solid wall is approximately 2.7 times greater than through the lightweight wall.

Example 1.4 *Figure 1.8 shows a section of a wall. Using the following data, calculate the thermal transmission (U) for the wall.*

Thermal conductivities

Brick	1.0 W/m °C
Aerated concrete	0.14 W/m °C
Plaster	0.7 W/m °C
Expanded polystyrene	0.04 W/m °C

Surface resistances

R_{si} External surface layer 0.12 m² °C/W

R_{so} Internal surface layer 0.08 m² °C/W

R_a Air space 0.18 m² °C/W

$$U = \frac{1}{R_{si} + R_{so} + R_a + \dfrac{L_1}{k_1} + \dfrac{L_2}{k_2} + \dfrac{L_3}{k_3} + \dfrac{L_4}{k_4}}$$

$$U = \frac{1}{0.12 + 0.08 + 0.18 + \dfrac{0.11}{1} + \dfrac{0.11}{0.14} + \dfrac{0.013}{0.7} + \dfrac{0.003}{0.04}}$$

$$U = \frac{1}{0.12 + 0.08 + 0.18 + 0.11 + 0.78 + 0.02 + 0.08}$$

$$U = \frac{1}{1.37}$$

$U = \underline{0.73 \text{ W/m}^2 \,^\circ\text{C}}$

The Institution of Heating and Ventilating Engineers Guide Book A gives a comprehensive table of *U* values for various types of structures. Table 1.5 gives *U* values for common types of structures.

Table 1.5 *U* values for common types of structures

Construction	U value (W/m² °C)		
	Sheltered	Normal	Severe
Walls (brickwork)			
220 mm solid brick wall unplastered	2.2	2.3	2.4
222 mm solid brick wall with 16 mm plaster	2.0	2.1	2.2
220 mm solid brick wall with 10 mm plasterboard on inside face	1.9	2.0	2.1
260 mm cavity wall (unventilated with 105 mm outer and inner leaves, with 16 mm plaster on inside face	1.4	1.5	1.6
Walls (brickwork)			
lightweight concrete block 260 mm cavity unventilated with 105 mm brick outer leaf, 100 mm lightweight concrete block inner leaf and 16 mm dense plaster on inside face	0.93	0.96	0.98
Walls, but with 13 mm expanded polystyrene board in cavity	0.69	0.70	0.71
Walls (lightweight concrete block)			
150 mm solid wall with 150 mm aerated concrete block, with tile hanging externally and with 16 mm plaster on inside face	0.95	0.97	1.0
Cavity wall (unventilated)			
with 76 mm aerated concrete block outer leaf, rendered externally 100 mm aerated concrete inner leaf and with 16 mm plaster on inside face	0.82	0.84	0.86
Concrete			
150 mm thick cast	3.2	3.5	3.9
200 mm thick cast	2.9	3.1	3.4
150 mm thick cast, with 50 mm woodwool slab permanent shuttering on inside face and 16 mm plaster	1.1	1.1	1.1

Table 1.5 – *continued*

Construction	U value (W/m² °C)		
	Sheltered	Normal	Severe
Roofs (flat or pitched)			
19 mm asphalt on 150 mm solid concrete	3.1	3.4	3.7
19 mm asphalt on 150 mm hollow tiles	2.1	2.2	2.3
19 mm asphalt on 13 mm cement and sand screed on 50 mm metal edge reinforced woodwool slabs on steel framing, with vapour barrier at inside	0.88	0.90	0.92
19 mm asphalt on 13 mm cement and sand screed, 50 mm woodwool slabs on timber joists and aluminium foil – backed 10 mm plasterboard ceiling, sealed to prevent moisture penetration	0.88	0.90	0.92
Roofs, but with 25 mm glass fibre insulation laid between joists	0.59	0.60	0.61
Tiles on battens, roofing felt and rafters, with roof space and aluminium foil-backed 10 mm plasterboard on ceiling joists	1.4	1.5	1.6
Tiles, but with boarding on rafters	1.4	1.5	1.6
Corrugated asbestos cement sheeting	5.3	6.1	7.2
Floors			
Suspended timber floor above ground:			
150 mm x 60 mm	—	0.14	—
150 mm x 30 mm	—	0.21	—
60 mm x 60 mm	—	0.16	—
60 mm x 30 mm	—	0.24	—
30 mm x 30 mm	—	0.28	—
30 mm x 15 mm	—	0.39	—
15 mm x 15 mm	—	0.45	—
7.5 mm x 7.5 mm	—	0.68	—
3 mm x 3 mm	—	1.05	—
Glass			
Single glazing	5.0	5.6	6.7
Double glazing with 20 mm air space	2.8	2.9	3.2
Double glazing with 12 mm air space	2.8	3.0	3.3
Double glazing with 6 mm air space	3.2	3.4	3.8
Double glazing with 3 mm air space	3.6	4.0	4.4
Triple glazing with 20 mm air space	1.9	2.0	2.1
Triple glazing with 12 mm air space	2.0	2.1	2.2
Triple glazing with 6 mm air space	2.3	2.5	2.6
Triple glazing with 3 mm air space	2.8	3.0	3.3

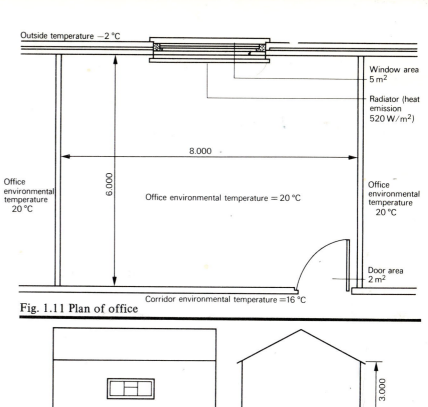

Fig. 1.11 Plan of office

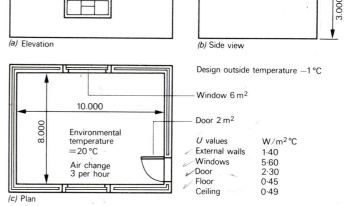

Fig. 1.12 Detached classroom

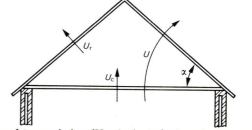

Fig. 1.13 Thermal transmission (U value) pitched roof

7

Heat loss through solid ground floor

A solid ground floor is exposed to the external air on one side near the wall and this is the point where the greatest heat loss will occur. The rate of heat loss decreases as the distance from the wall increases and therefore the rate of heat loss per unit area decreases as the floor area increases. The rate of heat loss through a solid ground floor therefore must be based on the floor area and the number of edges exposed. Figures 1.9 and 1.10 show how heat loss is lost through both narrow and wide solid ground floors respectively.

Table 1.6 shows U values for solid floors in contact with the earth.

Table 1.6

Dimensions of floor (metres)	Four exposed edges (W/m² °C)	Two exposed edges at right angles (W/m² °C)
150 × 60	0.11	0.06
150 × 30	0.18	0.10
60 × 60	0.15	0.08
60 × 30	0.21	0.12
30 × 30	0.26	0.15
30 × 15	0.36	0.21
15 × 15	0.45	0.26
15 × 7.5	0.62	0.36
7.5 × 7.5	0.76	0.45
3 × 3	1.47	1.07

Examples of heat loss calculations

In normal design situations the Institution of Heating and Ventilating Engineers Guide proposes two classifications of heat emitters: (a) convective, (b) radiant. In the former type, it is considered that the heat input to the space is at air temperature, while in the latter type it is considered to be at the environmental temperature. Heat emitters however do not fall rigidly into these classifications, but the guide suggests that heat emitters such as natural or forced convectors, overhead unit heaters, radiators and low temperature panel heaters should be considered as convective heaters. Heated floors and ceilings, high temperature radiant panels and strips should be considered as radiant heaters.

The formula heat loss = U value × area × temperature difference still applies.

Convective heating, using the environmental temperature concept

The following steps are taken:

(a) Calculate the heat loss through the various elements of the structure, using the difference between the inside and the outside environmental temperatures and the sum of these to produce ΣQ_f.

(b) Calculate the area of the entire enclosure to produce ΣA (this should include any partitions, floors, or ceilings, gaining heat from or losing heat to the adjacent rooms).

(c) Calculate $\Sigma Q_f / \Sigma A$ and from the following equation find the difference between the inside air temperature and inside environmental temperature.

$$t_{ai} - t_{ei} = \Sigma Q_f / 4.8 \, \Sigma A$$

where

t_{ai} = inside air temperature °C

t_{ei} = inside environmental temperature °C

ΣQ_f = rate of heat transfer through the building fabric W

ΣA = total area of room surfaces in m²

(d) Calculate the heat loss by ventilation. The heat loss due to infiltration may be found from the formula:

$$Q_v = pVC(t_{ai} - t_{ao})$$

where

p = density of air which may be taken as 1.2 kg/m³

V = infiltration rate m³/s

C = specific heat capacity of air, which may be taken as 1000 J/kg °C

t_{ai} = inside air temperature °C

t_{ao} = outside air temperature °C

Introducing the air infiltration rate N (air changes per hour) and room volume V (m³) the formula may be written,

$$Q_v = \frac{pNvC}{3600}(t_{ai} - t_{ao})$$

and substituting the above values for p and C,

$$Q_v = \frac{1.2 \times N \times V \times 1000}{3600}(t_{ai} - t_{ao})$$

$$Q_v = 0.33Nv(t_{ai} - t_{ao})$$

where

N = rate of air change per hour

v = volume of room in m³

t_{ai} = inside air temperature °C

t_{ao} = outside air temperature °C

(e) Find $\Sigma Q_f + \Sigma Q_v$ to give the total heat loss.

Worked examples on convective heating

Example 1.5. *Figure 1.11 shows the plan of an office on the second floor of a four-storey building. The other floors have the same construction and heating design conditions. From the data given below calculate the total rate of heat loss and the surface area of the radiator required.*

Office internal environmental temperature = 20 °C
Corridor environmental temperature = 16 °C
Design outside air temperature = −2 °C
Area of door = 2 m²
Area of window = 5 m²
Height of office = 3 m
Air change = 1 per hour
Ventilation allowance = 0.33 W/m³ °C

U *values*

External wall = 1.5 W/m² °C
Internal partition = 1.6 W/m² °C
Door = 2.3 W/m² °C
Window = 5.6 W/m² °C
Heat emission from radiator = 520 W/m²

Heat loss through the structure

Type of structure	Area (m²)	U value (W/m² °C)	Temperature difference (°C)	Total heat loss (watts)
External wall (less window)	19	1.5	22	627
Window	5	5.6	22	616
Internal partition (less door)	22	1.6	4	140.8
Door	2	2.3	4	18.4
Floor	48	—	— Nil	—
Ceiling	48	—	—	—
Other partition	36	—	—	—
ΣA	180		ΣQ_f	1402.2

Internal air temperature

$$t_{ai} - t_{ei} = \frac{\Sigma Q_f}{4.8 \, \Sigma A}$$

$$t_{ai} - t_{ei} = \frac{1402.2}{4.8 \times 180}$$

$$t_{ai} - t_{ei} = 1.6$$

$$t_{ai} = 20 + 1.6$$

$$t_{ai} = 21.6 \, °C$$

Heat loss by ventilation

$$Q_v = 0.33 Nv(t_{ai} - t_{ao})$$

$$Q_v = 0.33 \times 144[21.6 - (-1)]$$

$$Q_v = 0.33 \times 144 \times 22.6$$

$$Q_v = 1073.9W$$

Total heat loss Q_t

$$Q_t = \Sigma Q \, \Sigma Q_v$$

$$Q_t = 1402.2 + 1073.9$$

$$Q_t = 2476.1$$

Area of radiator

$$Area = \frac{\text{Total heat loss } W}{\text{Heat emission } W/m^2}$$

$$Area = \frac{2426}{520}$$

$$Area = 4.66 \text{ m}^2$$

Example 1.6 (Convective heating). *Figure 1.12 A, B, C shows the elevations and plan of a small detached classroom, to be heated by means of two convector heaters. Calculate the total heat loss for the classroom.*

Classroom internal environmental temperature 20 °C
Design outside air temperature 1 °C

Area of door	2 m²	
Area of windows	12 m²	
Height of classroom	3 m	
Air change	3 per hour	
Ventilation allowance	0.33 W/m³ °C	

U *values*

External wall 1.40 W/m² °C
Window 5.60 W/m² °C
Door 2.30 W/m² °C
Floor 0.45 W/m² °C
Ceiling 0.49 W/m² °C

Heat loss through structure

Type of structure	Area (m²)	U value (W/m² °C)	Temperature difference (°C)	Total heat loss (watts)
External wall (36 × 3) − (12 + 2) (less windows and doors)	94	1.4	21	2763.60
Window	12	5.6	21	1411.20
Door	2	2.3	21	96.60
Floor	80	0.45	21	756.00
Ceiling	80	0.49	21	823.20
ΣA	268		ΣQ_f	5850.60

Internal air temperature

$$t_{ai} - t_{ei} = \frac{\Sigma \, Q_f}{4.8 \, \Sigma \, A}$$

$$t_{ai} - t_{ei} = \frac{5850.6}{4.8 \times 268}$$

$$t_{ai} - t_{ei} = 4.548$$

$$t_{ai} = 20 + 4.5$$

$$t_{ai} = 24.5 \, ^\circ C$$

Heat loss by ventilation

$$Q_v = 0.33 N v (t_{ai} - t_{ao})$$

$$Q_v = 0.33 \times 3 \times 240 [24.5 - (-1)]$$

$$Q_v = 0.33 \times 3 \times 240 \times 25.5$$

$$Q_v = 6058.8 \, W$$

Total heat loss Q_t

$$Q_t = \Sigma \, Q_f + Q_v$$

$$Q_t = 5850.6 + 6058.8$$

$$Q_t = 11 \, 909.4 \, W$$

$$Q_t = 12 \, kW$$

Each convector would require a heat output of 6 kW.

Radiant heating

In order to find the total heat requirements, the following steps are necessary.

(a) Calculate the heat loss through the structure as for convective heating to produce $\Sigma \, Q_f$.

(b) Calculate the surface area of the entire enclosure, as for convective heating to produce $\Sigma \, A$.

(c) Find the ventilation conductance C_v from:

$$C_v = \frac{1}{0.33 N v} + \frac{1}{4.8 \, \Sigma \, A}$$

where N = number of air changes per hour

where v = volume of enclosure m².

(d) Calculate ventilation loss Q_v from:

$$Q_v = C_v (t_{ei} - t_{eo})$$

(e) Add the results of steps (a) and (d) to give the total heat loss.

Example 1.7 (Radiant heating). *Calculate the total heat loss for the classroom given in Example 1.6 but assume that in this case the heating is to be in the form of a heated ceiling.*

Note: Since a heated ceiling is to be employed, there is no need to consider heat loss from the classroom through the ceiling.

$$\Sigma \, Q_f = 5850.6 - 823.2$$

$$\Sigma \, Q_f = 5019.4 \, W$$

Also

$$\Sigma \, A = 268 \, m^2, \, v = 240, \, N = 3$$

Ventilation conductance

$$\frac{1}{C_v} = \frac{1}{0.33 \times N v} + \frac{1}{4.8 \, \Sigma \, A}$$

$$\frac{1}{C_v} = \frac{1}{0.33 \times 3 \times 240} + \frac{1}{4.8 \times 268}$$

$$\frac{1}{C_v} = 0.0042 + 0.000 \, 77$$

therefore

$$C_v = 201.2 \, W \, ^\circ C$$

Heat loss by ventilation

$$Q_v = C_v (t_{ei} - t_{eo})$$

$$Q_v = 201.2 [20 - (-1)]$$

$$Q_v = 201.2 \times 21$$

$$Q_v = 4225.2 \, W$$

Total heat loss Q_t

$$Q_t = \Sigma \, Q_f + \Sigma \, Q_v$$

$$Q_t = 5850 + 4225.2$$

$$Q_t = 10 \, 075.8$$

$$Q_t = 10 \, kW.$$

Pitched roofs

The U values for common types of pitched roofs can be found in the Institution of Heating and Ventilating Engineers Guide Book A. There may be occasions, however, when some special type of construction is to be used and a U value for the roof will have to be calculated. The following equation gives the method of calculating the U value for a pitched roof with a horizontal ceiling below. The values in the equation are shown in Figure 1.13.

$$U = \frac{U_r \times U_c}{U_r + U_c \cos \infty}$$

Example 1.8 *A pitched roof inclined at 40° has a horizontal ceiling below. If U_c and U_r are found to be 1.60 $W/m^2{}°C$ and 3.20 $W/m^2{}°C$ respectively, calculate the value of* U.

$$U = \frac{3.20 \times 1.60}{3.20 + (1.60 \times 0.766)}$$

$U = 1.157 \ W/m^2{}°C$

Example 1.9 *A pitched roof inclined at 45° has a horizontal ceiling below. If U_c and U_r are found to be 4.60 $W/m^2{}°C$ and 7.38 $W/m^2{}°C$ respectively, calculate the value of* U.

$$U = \frac{7.38 \times 4.60}{7.38 + (4.60 \times 707)}$$

$U = 3.19 \ W/m^2{}°C$

Alternatively, the U value for the total construction may be calculated by the use of the following formula:

$$U = \frac{1}{R_c + R_r \times \cos \infty}$$

where

R_c = thermal resistance of ceiling

R_r = thermal resistance of roof

The formula only applies where the roof space is unventilated: it can be used when considering the improvement made by adding an insulating lining to the roof by expressing the U value of the insulating material as a thermal resistance; the reciprocal of the total resistance gives the U value of the total construction.

Thermal insulation

Thermal insulation reduces the flow of heat through the structure of a building and its advantages may be summarised as follows:

1. The reduction in the size of the heating installation, resulting in the reduction in capital costs, fuel consumption and therefore running costs.

2. Saving in space for plant and fuel.
3. Reduction, or the complete elimination, of condensation problems.
4. Reduction of unsightly pattern staining and redecoration costs.
5. Improved comfort levels for the occupants of the building.
6. Reduced pre-heating time.
7. Reduced rate of heat gain from solar radiation and therefore reduction in size of cooling plant.

The inclusion of thermal insulation within the structure of a heated building can be regarded as an investment, from which an annual return on capital can be derived. The cost of insulation can also be largely offset by the saving in the cost of the heating installation.

The heat losses from a building having a compact layout are less than those from a straggling layout of the same floor area and volume. A room of square proportions on plan also has less heat losses than a room of rectangular proportions on plan. For example, a room measuring 6 m x 6 m has a perimeter of 24 m, while a room measuring 9 m x 4 m will have a perimeter of 26 m. The square room therefore will have less heat losses through the walls and floor perimeter than the rectangular room of the same floor area and volume.

Mandatory requirements

The revised thermal insulation standards 1975 made under the Building Regulations for England and Wales stipulate the following:

Building Regulations (Part F, Thermal Insulation)

Application of Part F

F1. 1. Subject to the provisions of paragraph (2) this Part shall apply to any building, or part of a building, which is intended to be used as a dwelling.
2. This Part shall not apply to any external wall, floor or roof of any part of a dwelling which consists of a shed or store entered from outside or of a garage, boathouse, conservatory or porch.

Interpretation of Part F

F2. 1. In this Part and in Schedule 11 —

'dwelling' means a house, flat or maisonette;
'partially ventilated space' means a space which —

(*a*) is either —

(i) a passage, stairway or other common space which is not part of, but adjoins, a dwelling; or
(ii) a part of a dwelling which consists of a shed or store or entered from outside or of a garage, boathouse, conservatory or porch; and

(*b*) is ventilated by means of permanent vents having an aggregate area not exceeding 30 per cent of its wall boundary area;
'perimeter walling' means those walls which together enclose all parts of a dwelling other than a partially ventilated space or a ventilated space;
'permanent vent' means an opening or duct which communicates

with the external air and is designed to allow the passage of air at all times;

'*U* value' means thermal transmittance coefficient, that is to say, the rate of heat transfer in watts through 1 m² of a structure when the combined radiant and air temperatures at each side of the structure differ by 1 °C and is expressed in W/m² °C;

'ventilated space' means a space which —

(a) is either —

(i) a passage, stairway or other common space which is not part of, but adjoins, a dwelling; or

(ii) a part of a dwelling which consists of a shed or store entered from outside or of a garage, boathouse, conservatory, or porch; and

(b) is ventilated by means of permanent vents having an aggregate area exceeding 30 per cent of its wall boundary area;

'wall boundary area' means the total superficial area of its walling, including any opening, bounding a partially ventilated space or a ventilated space; and

'window opening' means any structural opening which is provided for a window irrespective of its size and function or for a hinged or sliding door or panel having a glazed area of 2 m² or more.

2. For the purpose of this Part —

(a) unless the context otherwise requires, any reference to a dwelling is a reference solely to those parts of a dwelling which are enclosed by perimeter walling;

(b) any part of a roof which has a pitch of 70 per cent or more shall be treated as an external wall; and

(c) any floor which is so situated that its upper surface is exposed to the external air shall be treated as a roof in relation to that part of the building beneath it.

Maximum U *value of walls, floors, roofs and perimeter walling*

F3. 1. The *U* value of any part of a wall, floor or roof which encloses a dwelling and is described in this column (1) of the Table to this regulation (including surface finishes thereof and excluding any openings therein) shall not exceed the appropriate value specified in column (2) of that Table.

2. The calculated average *U* value of perimeter walling (including any opening therein) shall not exceed 1.8.

3. For the purposes of calculating the average *U* value of perimeter walling —

(a) the *U* value of any wall between a dwelling and another dwelling, or between a dwelling and an internal space which is within the same building and not ventilated by means of permanent vents, shall be assumed to be 0.5;

(b) the *U* value of a window opening shall be assumed to be 5.7 if it has single glazing and 2.8 if it has double glazing; and

(c) any other opening shall be assumed to have a *U* value equivalent to that of the wall in which it is situated.

Table to Regulation F3 (Maximum U *value of walls, floors and roofs)*

Element of building (1)	Maximum *U* value of any part of element (in W/m² °C) (2)
1. External wall	1.0
2. Wall between a dwelling and a ventilated space	1.0
3. Wall between a dwelling and a partially ventilated space	1.7
4. Wall between a dwelling and any part of an adjoining building to which Part F is not applicable	1.7
5. Wall or partition between a room and a roof space, including that space and the roof over that space	1.0
6. External wall adjacent to a roof space over a dwelling, including that space and any ceiling below that space	1.0
7. Floor between a dwelling and the external air	1.0
8. Floor between a dwelling and a ventilated space	1.0
9. Roof, including any ceiling to the roof or any roof space and any ceiling below that space	0.6

Deemed-to-satisfy provisions regarding thermal insulation

F4. 1. The requirements of regulation F3 (1) relating to the *U* value of any part of a wall, floor or roof shall be deemed to be satisfied if the wall, floor or roof is constructed in accordance with a specification contained in Part I, II, or III respectively of Schedule II.

2. The requirements of regulation F3 (2) relating to the average *U* value of perimeter walling shall be deemed to be satisfied if any one of the conditions prescribed in rule 2 of Part IV of Schedule II is satisfied.

The total window area must be included with the insulation for the walls. The Regulations exclude conservatories, garages, sheds, porches and boathouses. It has been calculated that a typical house, confirming to the insulation standards of the regulations and having central heating, controlled by the thermostats and programmer, has a heating load of about 80 watt per square metre of floor area.

Table 1.7

	Heat loss through structure (per cent)		
	Floor	Extended walls	Roof
Two-storey houses	10	60–65	25–30
Bungalows	15	40–45	40–45

Table 1.7 gives the relative structural heat losses through two-storey houses and bungalows.

The Building Standards (Scotland), Amendment 1975, governs the thermal insulation standards of domestic dwellings in Scotland.

The Amendment introduces a new Regulation which imposes a control aimed at reducing the risk of interstitial condensation in walls, which can be aggravated by the use of insulation without suitable vapour barriers, or with barriers that are incorrectly placed. The control requires the provision of a correctly placed vapour barrier in the wall structure.

In general, however, the English, Welsh and Scottish thermal regulations are similar.

The thermal insulation of Industrial Buildings Act 1957 requires that roofs of industrial buildings shall have a U value of 1.703 W/m^2 $^\circ$C, but exempts boiler houses, unheated buildings and buildings heated solely by a manufacturing or cleansing process. There are no mandatory requirements for offices, but the Standard of School Premises Regulations state that adequate insulation shall be provided.

Insulating materials

Insulating materials may be divided broadly into two groups:

1. Non-loadbearing.
2. Loadbearing.

Non-loadbearing materials generally have a low density and make use of still air cells and usually possess a greater thermal resistance than loadbearing materials. The rate of heat loss through modern non-loadbearing lightweight structures is less than the heat losses through heavier loadbearing structures and this has led to a wider use of lightweight prefabricated sections in buildings. Curtain wall structures use lightweight non-loadbearing panels between heavy frames of reinforced concrete, or structural steel. Some materials are organic in origin, or contain organic materials and the subject of fire spread must be considered in conjunction with that of insulation when proposing to use these materials. The hazards of smoke and toxic fumes produced in fires where foamed plastics are used for insulating materials must also be considered. The Building Regulations (England and Wales) Part E.15 covers the spread of fire in buildings and the local Fire Prevention Officer should also be consulted where any doubt exists regarding the choice of materials. When selecting a suitable insulating material, the following points must be considered:

1. The risk of spread of fire, or the production of toxic fumes when fixed.
2. Whether the material required, is non-loadbearing or loadbearing.
3. The cost and thickness in relation to the saving in fuel and capital cost of the heating installation.
4. The availability of the material and the ease of fixing.

Foamed plastics

Foamed plastics are the most effective of all materials used for thermal insulation and their thermal conductivities are as follows:

Foamed polyurethane	0.020–0.025 W/m $^\circ$C
Expanded polystyrene	0.029–0.04 W/m $^\circ$C
Foamed urea formaldehyde	0.030–0.036 W/m $^\circ$C

Cavity fill

Foamed urea formaldehyde resin is used for improving the thermal insulation of an external cavity wall (by about 50 per cent) (see Fig. 1.14). The foam is formed on site, by mixing the resin with a foaming agent, hardener and warm water in a cylinder. The foam formed in the cylinder is forced into the cavity under pressure, through 19 mm diameter holes at about 1 mm centres. Special care is required to ensure the complete filling of the cavity and specialist contractors are required to carry out the work. On new work, expanded polystyrene board 25 mm thick may be fixed to the inner leaf of the cavity, thus leaving 25 mm of an air cavity (see Fig. 1.15). Alternatively, on new work, the cavity may be filled with mineral wool quilt, or semi-rigid glass wool slab treated with a water-repellent binder.

Insulating boards or slab are used for insulating all various types of structures and these include the following types:

1. Wood wool slabs.
2. Compressed strawboard.
3. Insulating plasterboard.
4. Aluminium foil-backed plasterboard.
5. Asbestos boards.
6. Expanded polystyrene slabs.
7. Corkboard.
8. Insulating wood fibre board.
9. Cellular glass rigid slabs.
10. Semi-rigid slabs of glass fibre, treated with a water repellent.

Before the boards or slabs are fixed they must be conditioned for at least 24 hours, by exposing them on all sides to the same air temperature and humidity as would exist when they are fixed on site or otherwise distortion of the boards would occur after fixing.

Internal linings

Insulation board fixed to battens on the inside walls enables a room to warm up quickly and helps to prevent condensation when intermittent heating is used. The insulation will reduce the radiant heat loss from the body and therefore provides better conditions of thermal comfort. The insulation, however, should be considered with the risk of spread of fire since the temperature of the combustible materials inside the room will reach ignition point more quickly if a fire occurs.

Figures 1.16, 1.17 and 1.18 show the methods of fixing internal linings to brick and corrugated asbestos cement sheet walls.

Insulating plasters

Insulating plasters containing lightweight perlite and vermiculite as aggregates are produced ready for use and require only the addition of clean water before use. They have three times the insulating value and are only one-third the density of ordinary plasters.

Vapour barriers

In conditions where there is a risk of condensation, moisture vapour will rise through the structure and, being unable to disperse, may condense within the

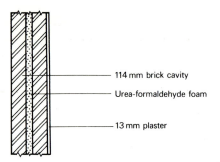

Fig. 1.14 Cavity wall with foam insulation

- 114 mm brick cavity
- Urea-formaldehyde foam
- 13 mm plaster

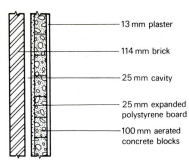

Fig. 1.15 Expanded polystyrene board insulation

- 13 mm plaster
- 114 mm brick
- 25 mm cavity
- 25 mm expanded polystyrene board
- 100 mm aerated concrete blocks

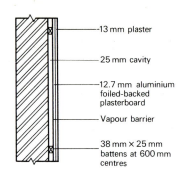

Fig. 1.16 Solid wall with interal lining

- 13 mm plaster
- 25 mm cavity
- 12.7 mm aluminium foiled-backed plasterboard
- Vapour barrier
- 38 mm × 25 mm battens at 600 mm centres

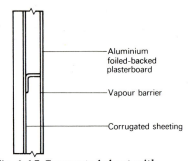

Fig. 1.17 Corrugated sheet with internal lining

- Aluminium foiled-backed plasterboard
- Vapour barrier
- Corrugated sheeting

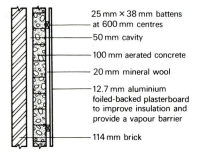

Fig 1.18 Cavity wall with internal lining

- 25 mm × 38 mm battens at 600 mm centres
- 50 mm cavity
- 100 mm aerated concrete
- 20 mm mineral wool
- 12.7 mm aluminium foiled-backed plasterboard to improve insulation and provide a vapour barrier
- 114 mm brick

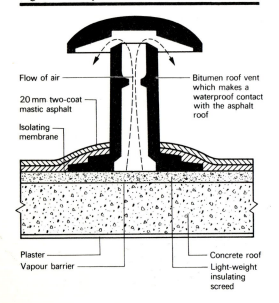

- Flow of air
- 20 mm two-coat mastic asphalt
- Isolating membrane
- Plaster
- Vapour barrier
- Bitumen roof vent which makes a waterproof contact with the asphalt roof
- Concrete roof
- Light-weight insulating screed

Fig. 1.19 Roof vent

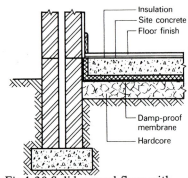

- Insulation
- Site concrete
- Floor finish
- Damp-proof membrane
- Hardcore

Fig 1.20 Solid ground floor with horizontal insulation

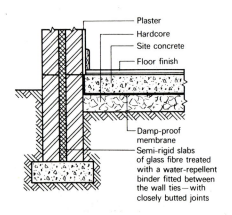

- Plaster
- Hardcore
- Site concrete
- Floor finish
- Damp-proof membrane
- Semi-rigid slabs of glass fibre treated with a water-repellent binder fitted between the wall ties — with closely butted joints

Fig. 1.21 Solid ground floor with edge insulation

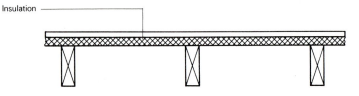

Insulation

Fig 1.22 Hollow timber ground floor with insulation board or slab above joists

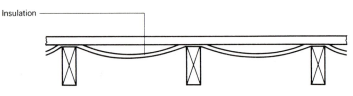

Insulation

Fig 1.23 Hollow timber ground floor with insulating quilt above joists

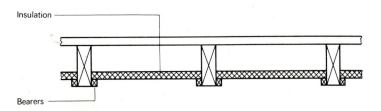

Insulation

Bearers

Fig. 1.24 Hollow timber ground floor with insulation board or slab between joists

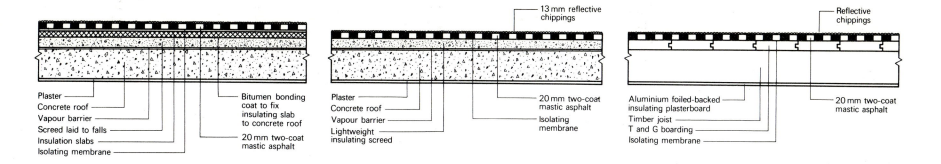

Plaster
Concrete roof
Vapour barrier
Screed laid to falls
Insulation slabs
Isolating membrane

Bitumen bonding coat to fix insulating slab to concrete roof
20 mm two-coat mastic asphalt

Fig. 1.25 Concrete roof with insulation slabs

13 mm reflective chippings

Plaster
Concrete roof
Vapour barrier
Lightweight insulating screed

20 mm two-coat mastic asphalt
Isolating membrane

Fig. 1.26 Concrete roof with insulating screed

Reflective chippings

Aluminium foiled-backed insulating plasterboard
Timber joist
T and G boarding
Isolating membrane

20 mm two-coat mastic asphalt

Fig 1.27 Timber roof with insulated ceiling

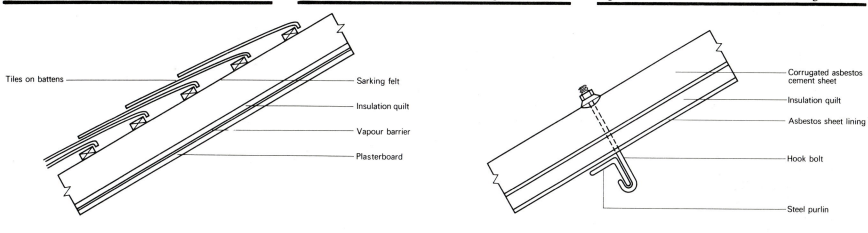

Tiles on battens

Sarking felt
Insulation quilt
Vapour barrier
Plasterboard

Fig. 1.28 Insulation of tiled roof

Corrugated asbestos cement sheet
Insulation quilt
Asbestos sheet lining
Hook bolt
Steel purlin

Fig. 1.29 Insulation of corrugated asbestos cement roof

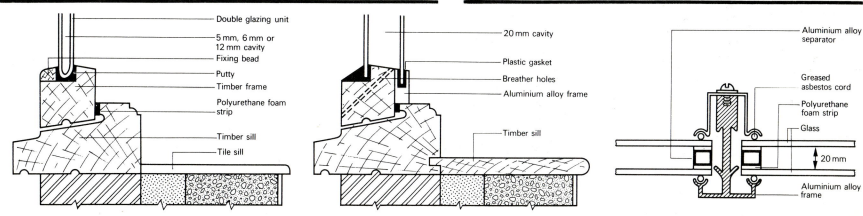

Double glazing unit
5 mm, 6 mm or 12 mm cavity
Fixing bead
Putty
Timber frame
Polyurethane foam strip
Timber sill
Tile sill

Fig. 1.30 Pilkington insulight double glazing units

20 mm cavity
Plastic gasket
Breather holes
Aluminium alloy frame
Timber sill

Fig. 1.31 Double glazing an existing window frame

Aluminium alloy separator
Greased asbestos cord
Polyurethane foam strip
Glass
20 mm
Aluminium alloy frame

Fig. 1.32 Aluminium alloy double glazing frame

structure and saturate the insulating material. If this occurs the insulating material will rapidly lose its insulating properties and if the material is organic in origin it will decompose. In such situations a suitable vapour barrier should be included in the structure and should be inserted on the warmest side of the structure.

Roof vents

In concrete roofs moisture may be present in the screed below the mastic asphalt or felt covering. During hot weather this moisture may vaporise and cause an upward pressure, resulting in the lifting of the roof covering. To prevent this occurrence roof ventilators are fixed to relieve this pressure. It is usual to fix one ventilator to every 20 m^2 of roof area (see Fig. 1.19).

Solid ground floors

Since the greatest heat loss is around the perimeter of the floor, it is essential to provide insulation at this location. For wide floors, a horizontal strip of insulation, may be extended 1 m from the wall and a vertical strip of insulation material should be extended through the thickness of the floor slab, around all exposed edges (see Fig. 1.20). For narrow floors, the horizontal strip insulation may be carried under the entire floor slab, with vertical strip insulation material extended through the thickness of the floor slab, as before. If the cavity is to be filled with semi-rigid slabs of glass fibre, specially treated with a water-repellent binder, horizontal strip insulation may be omitted (see Fig. 1.21). The slabs are available in three thicknesses, 50, 65 and 75 mm, and they provide a moisture barrier between the outer and inner leaf of the cavity wall. If required, a horizontal strip 1 m from the wall may also be included, to provide additional insulation for the floor.

Suspended timber ground floor

A suspended ground floor above an enclosed air space is exposed to air on both sides, but the air temperature below the floor is higher than the external air because the ventilation rate below the floor is very low. Additional thermal insulation of a suspended timber ground floor may be provided either in the form of a continuous layer of semi-rigid or flexible material laid over the joists (see Figs 1.22 and 1.23). Alternatively, a semi-rigid material may be laid between the joists (see Fig. 1.24). A vapour barrier should never be included in suspended timber ground floors as this would prevent the escape of water above and evaporation, which could lead to the dangerous conditions conductive to fungal attack on the timber floor.

Flat roofs

These are usually constructed of reinforced concrete or timber and may be covered with mastic asphalt, roofing felt, lead, copper, aluminium or zinc. To prevent radiant heat gain, mastic asphalt and felt roofs have a layer of reflective chippings spread over the external surface. Figure 1.25 shows a section through a concrete flat roof, insulated with mineral wool slabs and Fig. 1.26 shows an alternative method of insulating a concrete flat roof with an aerated concrete screed. Figure 1.27 shows a section through a timber roof, insulated with aluminium foil-backed plasterboard. The aluminium foil acts as both a thermal insulator and a vapour barrier.

Environmental temperature (t_e) As mentioned earlier,

this is a balanced mean, between the mean radiant temperature and the air temperature. It may be evaluated approximately from the following formula:

$$t_{ei} = \frac{2}{3} t_{ri} + \frac{1}{3} t_{ai}$$

where t_{ri} = mean radiant temperature of all room surfaces (°C)

t_{ai} = inside air temperature (°C)

The approximate value of the mean radiant temperature may be found by totalling the products of the various areas and the surrounding surfaces and dividing this total by the sum of the areas; i.e.,

$$t_{ri} = \frac{a_1 t_1 + a_2 t_2 + a_3 t_3 + \ldots}{a_1 + a_2 + a_3 + \ldots}$$

where

t_1, t_2, t_3 = surface temperature (°C)
a_1, a_2, a_3 = surface areas (m^2)

Table 1.8 provides a summary of recommended environmental temperatures.

Table 1.8 Summary of recommended internal (winter) environmental temperatures

	t_{ei} (°C)		t_{ei} (°C)
Art galleries	20	Laboratories	20
Assembly halls	18	Law courts	20
Bars	18	Libraries	20
Canteens	20		
Churches	18	*Offices*	
		General	20
Factories		Private	20
Sedentary work	19	Police stations	18
Light work	16	Restaurants	18
Heavy work	13		
		Hotels	
Flats and houses		Bedrooms (standard)	22
Living rooms	21	Bedrooms (luxury)	24
Bedrooms	18	Public rooms	21
Bathrooms	22		
Entrance hall	16	*Schools and colleges*	
		Classrooms	18
Hospitals		Lecture rooms	18
Corridors	16		
Offices	20	*Shops*	
Operating theatre	18–21	Small	18
Wards	18	Large	18
Sports pavilions	21	Swimming baths	
Warehouses	16	Changing rooms	22
		Bath hall	26

Example 1.10 *Using the following data, calculate the environmental temperature of a room in which the air temperature is 22 °C.*

Fabric	Area (m²)	Surface temperature (°C)
Floor	200	18
Walls	180	20
Windows	60	10
Ceiling	200	16

$$t_{ri} = \frac{(200 \times 18) + (180 \times 20) + (60 \times 10) + (200 \times 16)}{200 + 180 + 60 + 20}$$

$$= \frac{3600 + 3600 + 600 + 3200}{460}$$

$$= \frac{11\,000}{460}$$

$$= 23.9\,°C$$

The environmental temperature may be found as follows:

$$t_{ei} = \frac{2}{3}\,t_{ri} + \frac{1}{3}\,t_{ai}$$

$$= \left(\frac{2}{3}\right) \times 23.9 + \left(\frac{1}{3}\right) \times 22$$

$$= 15.933 + 7.333$$

$$= 23.266\,°C$$

$$= 23.3\,°C \text{ (approx.)}$$

Whether the calculation is carried out by the environmental-temperature concept or the internal air-temperature method, the following formula for the heat loss through the structure is used:

heat loss in watts = area (m²) × *U* value × temperature difference

Example 1.11 *Compare the heat losses in watts through two walls having U values of 1.5 W/m² °C and 0.70 W/m² °C respectively, when the area of each wall is 200 m². The internal environmental temperature may be taken as 20 °C when the outside air temperature is − 1 °C.*

heat loss = area × *U* × $(t_{ei} - t_{ao})$

where   area = m²

         U = thermal transmission (W/m² °C)

        t_{ei} = environmental temperature inside (°C)

        t_{ao} = air temperature outside (°C)

For walls having a *U* value of 1.50 W/m² °C,

     heat loss = 200 × 1.50 × (20 − (−1))

            = 200 × 1.50 × 21

            = 6300 watts

For walls having a *U* value of 0.70 W/m² °C,

     heat loss = 200 × 0.70 × 21

            = 2940 watts

difference
in heat loss = 6300 − 2940 = 3360 watts

percentage
    saving = $\frac{2940}{6300} \times \frac{100}{1}$

        = 46.66   per cent

Heat emission from a radiator

Manufacturers usually give the radiator heat emission for a temperature differential between the mean water temperature and the ambient, or surrounding, air temperature of 55.6 °C.

For a temperature differential of other than 55.6 °C, the radiator heat emission may be found from the following formula:

$$E_2 = E_1 \times \left[\frac{\Delta t}{55.6}\right]^{1.3}$$

where E_2 = heat emission in W/m² for the new temperature differential

        E_1 = heat emission in W/m² for a 55.6 °C differential

        Δt = the new temperature differential in °C

Example 1.12 *Calculate the heat emission from a radiator in W/m² when the temperatures of the flow and return waters are 80 °C and 70 °C respectively, when the ambient air temperature is 18 °C. The beat emission for the radiator given by the manufacturers for a temperature differential of 55.6 °C is 520 W/m².*

$$\Delta t = \left[\frac{80 + 70}{2}\right] - 18$$

$$\Delta = 75 - 18$$
$$= 57\,°C$$

Substituting in the formula,

$$E_2 = 520 \times \left[\frac{57}{55.6}\right]^{1.3}$$

$$= 520 \times 1.025^{1.3}$$

$$= 520 \times 1.030 = 535.6\,W/m²$$

Heat emission from a convector

When considering the heat emission from natural convectors for temperature differentials other than 55.6 °C, the following formula may be used:

$$E_2 = E_1 \times \left[\frac{\Delta t}{55.6} \right]^{1.5}$$

If the heat emitter in example 1.12 is a natural convector, the heat emission would be:

$$E_2 = 520 \left[\frac{57}{55.6} \right]^{1.5}$$
$$= 539.76 \text{ W/m}^2$$

Example 1.13   *Figure 1.33 shows the plan of a two-bedroomed bungalow. Using the following factors, calculate the heat losses in kW and the required radiator areas in m². The bungalow is to be well insulated and the internal air-temperature method of calculating the heat loss through the external walls may therefore be used.*

Factors

1.	Design temperature (°C)		Air change per hr
	Dining room	22	2
	Lounge	22	2
	Kitchen	18	2
	Bedrooms	18	1½
	Bathroom	22	2
	Hall	18	1½
	Corridor	18	1½

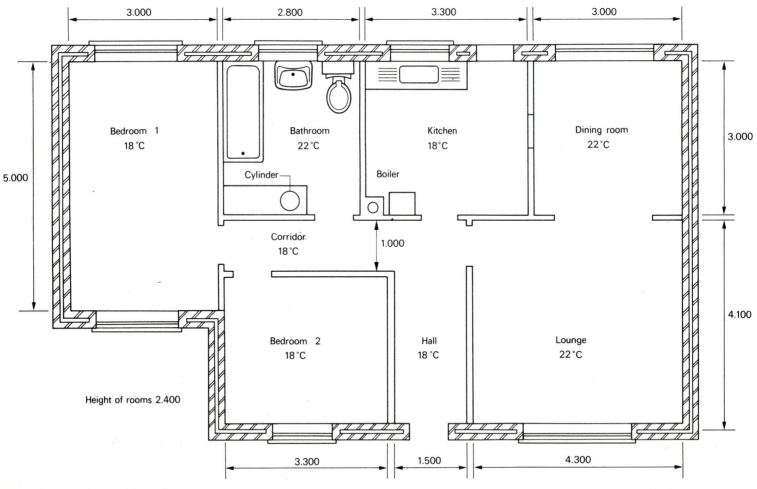

Fig. 1.33 Plan of two-bedroomed bungalow

2. U values (W/m² °C)

Floors	0.30
External walls	0.70
Partition walls	2.20
Windows	2.80
Doors	2.40
Ceilings	0.40

3. External air temperature $-1\,°C$.

4. Single-panel steel radiators having a heat emission of 530 W/m².

5. Window areas (m²)

Dining room	2.0 × 1.8
Lounge	1.7 × 1.3
Kitchen	1.2 × 1.0
Bathroom	1.2 × 1.0
Bedroom one (2)	1.7 × 1.2
Bedroom two	1.7 × 1.2

6. Areas of doors (m²) 1.8 × 0.76

7. Air volumes

		m³
Dining room	3.0 × 3.0 × 2.4	21.60
Lounge	4.3 × 4.1 × 2.4	42.312
Hall	3.2 × 1.5 × 2.4	11.52
Corridor	4.8 × 1.0 × 2.4	11.52
Kitchen	3.3 × 3.0 × 2.4	23.76
Bathroom	3.0 × 2.8 × 2.4	20.16
Bedroom one	5.0 × 3.0 × 2.4	36.00
Bedroom two	3.3 × 3.09 × 2.4	24.47

Building Regulations (Section F: Thermal Insulation)

Before calculating the heat losses from the bungalow a check must be made to find if the construction satisfies the regulations.

Step 1

Calculate the areas of different U values and from the following formula find the average U-value (Ua) of solid wall and window openings.

$$Ua = \frac{5.7\,A_1 + 2.8\,A_2 + Uf Af}{A_1 + A_2 + Af}$$

Where

A_1 = Total area of single glazed openings
A_2 = Total area of double glazed openings
Af = Area of walling with required U-value not exceeding 1.0
Uf = Actual U-value of wall required to have a U-value not exceeding 1.0

Values required

$A_1 = 0$
$A_2 = 14.33$ m²
$Af = 96.816$ (including doors)
$Uf = 0.7$

$$\therefore \quad Ua = \frac{(2.8 \times 14.33) + (0.7 \times 96.816)}{14.33 + 96.816}$$

$Ua = 0.97$ approx.

Step 2

Calculate the average U-value for the perimeter wall, using the following formula:

$$U = \frac{Aa\,Ua + Ab\,Ub + Ac\,Uc}{Aa + Ab + Ac} \quad : \text{not to exceed 1.8}$$

Where

Aa = Area of wall required to have a U-value not exceeding 1.0 and includes area of window openings
Ab = Area of wall required to have a U-value not exceeding 1.7
Ac = Area of wall assumed to have a U-value not exceeding 0.5
Ua is found from the previous formula
Ub = U-value of wall with required U-value not exceeding 1.7
Uc = U-value of wall with required U-value not exceeding 0.5 (the U-value for walls between dwellings)

Values required

$Aa = 111.146$ m²
$Ua = 0.97$

$$\therefore \quad U = \frac{111.146 \times 0.97}{111.146}$$

$U = 0.97$

This is below the maximum U-value of 1.8 W/m² °C and is therefore satisfactory.

Area of windows

It may sometimes be necessary to find the maximum area of double or single glazing necessary to ensure that the average U-value of the perimeter wall does not exceed 1.8 W/m² °C.

If in the example given for the bungalow the U-value for the cavity wall is to be 0.9 W/m² °C and single glazing is to be used having a U-value of 5.7 W/m² °C, the maximum area of single glazing would be as follows:

$$U = \frac{Aa \left[\dfrac{5.7A_1 + Uf\,Af}{A_1 + Af} \right]}{Aa}$$

$$1.8 = \cfrac{113.186 \left[\cfrac{5.7A_1 + 0.9 \times 96.816}{A_1 + 96.816}\right]}{113.186}$$

$$1.8 = \frac{5.7A_1 + 0.9 \times 96.816}{A_1 + 96.816}$$

$$1.8\,[A_1 + 96.816] = 5.7A_1 + 87.134$$

$$1.8\,A_1 + 1.8 \times 96.816 = 5.7A_1 + 87.134$$

$$174.268 - 87.134 = 5.7A_1 - 1.8\,A_1$$

$$87.134 = 3.9\,A_1$$

$$A_1 = \frac{87.134}{3.9}$$

Maximum area of windows = 22.342 m^2

In order to calculate the heat loss through the structure a suitable table may be used as in Tables 1.9 -1.16.

Table 1.9 Dining room

Type of structure	Dimensions (m)	Area (m^2)	U value (W/m^2 $^\circ$C)	Temperature difference ($^\circ$C)	Heat loss (W)
External wall	9 × 2.4 deduct 2 × 1.8 (window)	18	0.7	23	289.80
Window	2 × 1.8	3.6	2.8	23	231.84
Partition wall	3 × 2.4	7.2	2.2	4	63.36
Floor	3 × 3	9	0.3	23	62.10
Ceiling	3 × 3	9	0.4	23	82.80
	heat loss through structure			Total	729.90

Table 1.10 Lounge

Type of structure	Dimensions (m)	Area (m^2)	U value (W/m^2 $^\circ$C)	Temperature difference ($^\circ$C)	Heat loss (W)
External wall	8.4 × 2.4 deduct 1.7 × 1.3	17.95	0.70	23	288.995
Window	1.7 × 1.3	2.21	2.80	23	142.324
Partition wall	4.1 × 2.4 deduct 1.8 × 0.76	8.472	2.20	4	74.554
Door	1.8 × 0.76	1.368	2.40	4	13.133
Floor	4.3 × 4.1	17.63	0.30	23	121.647
Ceiling	4.3 × 4.1	17.63	0.40	23	162.196
	heat loss through structure			Total	802.849

Dining room

ventilation losses
$$Q_v = 0.333\,N\,V\,(t_{ai} - t_{ao})$$
$$= 0.333 \times 2 \times 21.6 \times 23$$
$$= 330.8688$$

total heat losses = 729.90 + 330.87
$$= 1060.77 \text{ watts}$$

area of radiator = $\dfrac{1060.77}{530}$ = 2 m^2 (approx.)

Lounge

ventilation losses
$$Q_v = 0.333\,N\,V\,(t_{ai} - t_{ao})$$
$$= 0.333 \times 2 \times 42.312 \times 23$$
$$= 648.135$$

total heat losses = 802.849 + 648.135
$$= 1450.984 \text{ watts}$$

area of radiator = $\dfrac{1450.984}{530}$
$$= 2.74 \text{ m}^2 \text{ (approx.)}$$

Table 1.11 Hall and corridor (heat loss)

Type of structure	Dimensions (m)	Area (m^2)	U value (W/m^2 $^\circ$C)	Temperature difference ($^\circ$C)	Heat loss (W)
External wall	1.5 × 2.4 deduct 1.8 × 0.76	2.232	0.70	19	29.686
Door	1.8 × 0.76	1.368	2.40	19	62.380
Floor	(4.1 × 1.5) add	9.45	0.30	19	53.865
Ceiling	3.3 × 1.0	2.85	0.40	19	21.660
	heat loss through structure			Total	167.591

Table 1.12 Hall and corridor (heat gains)

Type of structure	Dimensions (m)	Area (m^2)	U value (W/m^2 $^\circ$C)	Temperature difference ($^\circ$C)	Heat loss (W)
Partition	(2.8 + 4.1) × 2.4 deduct 2 doors 1.8 × 0.76	13.824	2.20	4	121.651
Doors	2 × 1.8 × 0.76	2.736	2.40	4	26.265
	heat gains through structure			Total	147.916

Hall and corridor

In order to find the effective heat loss through the structure, the heat gains must be deducted from the heat losses.

$$\text{effective heat loss} = 167.591 - 147.916$$
$$= 19.675 \text{ watts}$$

ventilation losses

$$Q_v = 0.333\, N\, V\, (t_{ai} - t_{ao})$$
$$= 0.333 \times 1.5 \times (2 \times 11.52) \times (18 - (-1))$$
$$= 0.333 \times 1.5 \times 23.04 \times 19$$
$$= 218.66$$

$$\text{total heat losses} = 19.675 + 218.66$$
$$= 238.335 \text{ watts}$$

$$\text{area of radiator} = \frac{238.335}{530}$$
$$= 0.449$$
$$= 0.5 \text{ m}^2 \text{ (approx.)}$$

Table 1.13 Bathroom (heat losses)

Type of structure	Dimensions (m)	Area (m²)	U values (W/m²°C)	Temperature difference (°C)	Heat loss (W)
External wall	2.8 × 2.4 deduct 1.2 × 1.0	5.52	0.70	23	88.872
Window	1.2 × 1.0	1.2	2.80	23	77.280
Partition walls	11.6 × 2.4 deduct 7.8 × 0.76	26.472	2.20	4	232.954
Door	1.8 × 0.76	1.368	2.40	4	13.133
Floor	3.0 × 2.8	8.4	0.30	23	57.96
Ceiling	3.0 × 2.8	8.4	0.40	23	77.28
	heat loss through structure			Total	547.479

Bathroom

ventilation losses

$$Q_v = 0.333\, N\, V\, (t_{ai} - t_{ao})$$
$$= 0.333 \times 2 \times 20.16 \times 23$$
$$= 308.80$$

$$\text{total heat losses} = 547.479 + 308.8$$
$$= 856.28 \text{ watts}$$

$$\text{area of radiator} = \frac{856.28}{530}$$

$$= 1.616$$
$$= 1.6 \text{ m}^2 \text{ (approx.)}$$

Table 1.14 Bathroom 1

Type of structure	Dimensions (m)	Area (m²)	U value (W/m²°C)	Temperature difference (°C)	Heat loss (W)
External walls	11 × 2.4 deduct 2 (1.7 × 1.2)	22.32	0.70	19	296.856
Windows	2 (1.7 × 1.2)	4.08	2.80	19	217.056
Floor	5 × 3	15.0	0.40	19	85.500
Ceiling	5 × 3	15.0	0.40	19	114.000
					713.412
Heat gain Partition wall	3 × 2.4	7.2	2.2	4	63.360
	heat loss through structure			Total	650.052

Bedroom 1

ventilation losses

$$Q_v = 0.333\, N\, V\, (t_{ai} - t_{ao})$$
$$= 0.333 \times 1.5 \times 36 \times 19$$
$$= 341.658$$

$$\text{total heat losses} = 650.052 + 341.658$$
$$= 991.710 \text{ watts}$$

$$\text{area of radiators} = \frac{991.710}{530}$$
$$= 1.871$$

Use two radiators, 1 m² each.

Kitchen

ventilation losses

$$Q_v = 0.333\, N\, V\, (t_{ai} - t_{ao})$$
$$= 0.333 \times 2 \times 23.76 \times 19$$
$$= 300.659$$

$$\text{total heat loss} = 25.394 + 300.659$$
$$= 326.053 \text{ watts}$$

The heat losses from the boiler would adequately cover the above heat loss and therefore a radiator in the kitchen is not normally required. The heat loss from the kitchen, however, must be added to the heat losses from the other rooms in order to find the boiler power.

Table 1.15 Kitchen

Type of structure	Dimensions (m)	Area (m²)	U value (W/m²°C)	Temperature difference (°C)	Heat loss (W)
External walls	3.3 × 2.4 deduct 1.2 × 1.0 + 1.8 × 0.76	5.352	0.70	19	71.182
Window	1.2 × 1.0	1.20	2.80	19	6.384
Door	1.8 × 0.76	1.368	2.40	19	6.238
Floor	3.3 × 3.0	9.9	0.30	19	56.430
Ceiling	3.3 × 3.0	9.9	0.40	19	75.240
					215.474

Heat gain

Type of structure	Dimensions (m)	Area (m²)	U value	Temperature difference	Heat loss
Partition walls	9.0 × 2.4	21.6	2.20	4	190.080
heat loss through structure				Total	25.394

Table 1.16 Bedroom 2

Type of structure	Dimensions (m)	Area (m²)	U value (W/m²°C)	Temperature difference (°C)	Heat loss (W)
External walls	(3.3 + 1.9) × 2.4 deduct 1.7 × 1.2	10.44	0.7	19	138.852
Window	1.7 × 1.2	2.04	2.8	19	108.528
Floor	3.3 × 3.09	10.197	0.3	19	58.123
Ceiling	3.3 × 3.09	10.197	0.4	19	77.497
heat loss through structure				Total	383.000

Bedroom 2

ventilation losses

$$Q_v = 0.333\, N\, V\, (t_{ai} - t_{ao})$$
$$= 0.333 \times 1.5 \times 24.47 \times 19$$
$$= 232.233$$

total heat losses $= 383.000 + 232.233$
$$= 615.233 \text{ watts}$$

area of radiator $= \dfrac{615.233}{530}$
$$= 1.16$$
$$= 1.2 \text{ m}^2 \text{ (approx.)}$$

Total heat losses (watts)

Dining room	1060.77
Lounge	1450.974
Hall and corridor	238.335
Bathroom	856.280
Bedroom 1	991.715
Bedroom 2	615.233
Kitchen	326.053
	5539.360

Boiler power

To this heat loss, an allowance of 20 per cent may be added for the heat losses from the pipes and an allowance of 3000 watts for the hot-water supply.

boiler power = 5539.360 + 1107.872 + 3000
boiler power = 9647.232
= 10 kW (approx.)

Questions

1. Define the following terms: (*a*) thermal conductivity; (*b*) thermal resistivity; (*c*) thermal conductance; (*d*) thermal resistance; (*e*) thermal transmission.

2. One wall of a building consists of 25 mm thickness of cedar boardings; 76 mm thickness of glass wool; and 13 mm thickness of plasterboard. Using the following values, calculate the thermal transmission (U) for the wall.

Thermal conductivity

Cedar wood	0.14 W/m°C
Glass wool	0.042 W/m°C
Plasterboard	0.58 W/m°C

Thermal resistances

Inside surface	0.123 m²°C/W
Outside surface	0.055 m²°C/W

Answer: 0.457 W/m²°C

3. If the thermal transmission for a wall is 0.74 W/m²°C, calculate the heat loss in watts through the wall having an area of 36 m².

Answer: 2.664 watts

4. Figure 1.34 shows the plan of an office having an internal environmental temperature of 20°C when the external air temperature is −1°C. Using the environmental concept for convective heating and an air change of 2 per hour, calculate the rating of the convector heaters in kW. Use the following values:

(*a*) area of each window 4 m².
(*b*) area of each door 1.4 m².

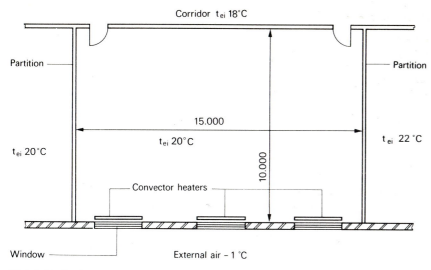

Corridor t_{ei} 18°C

Partition

Partition

15.000

t_{ei} 20°C

t_{ei} 22°C

t_{ei} 20°C

10.000

Convector heaters

Window

External air −1 °C

Fig. 1.34 Plan of an office

(c) **U values (W/m² °C)**

Floor	1.40
External walls	0.96
Partition walls	2.2
Windows	5.7
Doors	2.4
Ceiling	0.9

Answer: 4.8 kW (approx.). Use three heaters, 1.6 kW each.

5. Using the following data, calculate the environmental temperature of a room in which the air temperature is 22 °C:

Fabric	Area (m²)	Surface temperature (°C)
Floor	150	20
Walls	150	20
Windows	30	8
Ceiling	150	18

Answer: 19.73 °C

6. A pitched roof inclined at 30° has a horizontal ceiling below. If the U values for the roof and ceiling are found to be 2.80 W/m² °C and 0.7 W/m² °C respectively, calculate the combined U value for the ceiling and roof. Does this value satisfy the Building Regulations?

Answers: 0.575 W/m² °C; yes, it is below 0.6

7. Calculate the heat emission from a radiator in W/m² when the temperatures of the flow and return waters are 82 °C and 71 °C respectively, when the ambient air temperature is 16 °C. The heat emission for the radiator given by the manufacturers for a temperature differential of 55.6 °C is 530 W/m².

Answer: 591.5 W/m² (approx.)

8. Calculate the heat lost by ventilation in a room measuring 15 m by 10 m by 3 m, when the rate of air change is six per hour and the inside and outside air temperatures are 24 °C and −10 °C respectively.

Answer: 30.569 kW (approx.)

Chapter 2

Temperature drop through structures and condensation

Temperature drop through structures

To find the temperature distribution through a structure, it is necessary to know the following values:

1. the thermal resistances of the materials forming the structure;
2. temperature on either side of the structure.

Example 2.1 (see Fig. 2.1). *Calculate the U value and find the internal and external surface temperatures of a 105 mm thick solid brick wall when the internal and external air temperatures are 22°C and 2°C respectively.*
  Use the following data:

external surface resistance	*= 0.053 m² °C/W*
internal surface resistance	*= 0.123 m² °C/W*
thermal conductivity of brick	*= 1.20 W/m°C*

Thermal resistance of brick wall =

Thermal·resistance of brick wall = $\dfrac{L}{k} = \dfrac{0.105}{1.20} = 0.0875$

$$U \text{ value} = \frac{1}{R_e + R_i + R_b}$$

$$= \frac{1}{0.053 + 0.123 + 0.0875}$$

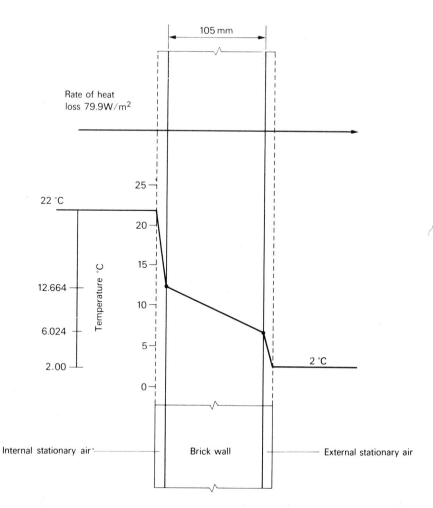

Fig. 2.1 Graph of wall temperatures

$$= \frac{1}{0.2635}$$

$$= 3.796 \text{ W/m}^2 \text{ °C}$$

rate of heat los = $U \times$ temperature difference

$$= 3.796 \times (22 - 2)$$

$$= 75.9 \text{ W/m}^2$$

If 75.9 W/m² is passing through the whole structure, it will pass through each separate resistance.

Internal surface of brick wall

rate of heat loss = $U \times$ temperature difference

$$= \frac{1}{R} \times \text{ temperature difference}$$

25

$$75.9 = \frac{1}{0.123} \times \text{temperature difference}$$

$\therefore$ temperature

difference $= 75.9 \times 0.123 = 9.336\,^\circ\text{C}$

temperature of
stationary air or
internal surface $= 22 - 9.336 = 12.664\,^\circ\text{C}$

External surface of brick wall

$$75.9 = \frac{1}{0.0875} \times \text{temperature difference}$$

$\therefore$ temperature

difference $= 75.9 \times 0.0875 = 6.64\,^\circ\text{C}$

temperature of
outer surface $= 12.664 - 6.64 = 6.024\,^\circ\text{C}$

External air (for checking purposes)

$$75.9 = \frac{1}{0.053} \times \text{temperature difference}$$

$\therefore$ temperature

difference $= 75.9 \times 0.053 = 4.023\,^\circ\text{C}$

temperature of
external air $= 6.024 - 4.023 = 2\,^\circ\text{C}$

Condensation

The terms used in problems of condensation are:

1. **Temporary condensation**   This is condensation occurring on the internal surfaces when a sudden rise in the air temperature causes air in contact with surfaces to be temporarily at a much higher temperature than the surfaces. If the surfaces are below the dewpoint of the air, condensation will occur.
2. **Permanent condensation**   In poorly insulated buildings the inside surfaces are at relatively low temperatures and if the internal air is comparatively humid, the internal surfaces may be at all times below the dewpoint temperature of the air. In these conditions condensation will be permanent.
3. **Interstitial condensation**   This is condensation occurring within a structure. It is possible for interstitial condensation to occur although condensation on the internal surfaces is absent.

Example 2.2 (see Fig. 2.2).   *If the brick wall in example 2.1 has 25 mm thickness of wood-wool slabs added to the internal surface, calculate the U value and find if interstitial condensation will occur if the dewpoint temperature of the air is 5°C. The thermal conductivity of wood-wool is 0.09 W/m² °C.*

thermal resistance of wood-wool slabs $= \dfrac{0.025}{0.09} = 0.278$

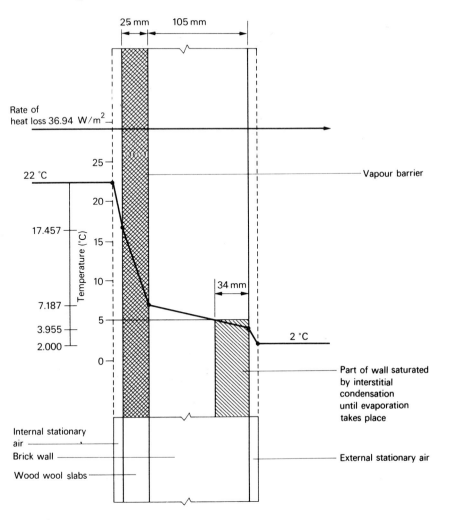

Fig. 2.2   Graph of wall temperatures and interstitial condensation

$$U \text{ value for wall} = \frac{1}{0.053 + 0.123 + 0.0875 + 0.278}$$

$$= \frac{1}{0.5415}$$

$$= 1.847 \text{ W/m}^2\,^\circ\text{C}$$

rate of heat loss for 1 m² of structure

$$= U \text{ value} \times \text{temperature difference}$$

$$= 1.847 \times (22 - 2)$$

$$= 36.94 \text{ watts}$$

Internal surface

$$\text{rate of heat loss} = U \times \text{temperature difference}$$

$$= \frac{1}{R} \times \text{temperature difference}$$

$$36.94 = \frac{1}{0.123} \times \text{temperature difference}$$

$$\text{temperature difference} = 36.94 \times 0.123$$
$$= 4.543\,^{\circ}C$$

$$\begin{aligned}\text{temperature of stationary} \\ \text{air or internal surface} &= 22 - 4.543 \\ &= 17.457\,^{\circ}C\end{aligned}$$

Inner face of brick wall

$$36.94 = \frac{1}{0.278} \times \text{temperature difference}$$

$$\text{temperature difference} = 36{:}94 \times 0.278$$
$$= 10.27\,^{\circ}C$$

$$\text{temperature at inner face} = 17.457 - 10.27$$
$$= 7.187\,^{\circ}C$$

Outer face of brick wall

$$36.94 = \frac{1}{0.0875} \times \text{temperature difference}$$

$$\text{temperature difference} = 36.94 \times 0.0875$$
$$= 3.232\,^{\circ}C$$

$$\text{temperature at outer face} = 7.187 - 3.232$$
$$= 3.955\,^{\circ}C$$

Note: This is below dewpoint temperature and condensation will occur between $5\,^{\circ}C$ and $3.955\,^{\circ}C$.

Temperature of external air

$$36.94 = \frac{1}{0.053} \times \text{temperature difference}$$

$$\text{temperature difference} = 36.94 \times 0.053$$
$$= 1.958\,^{\circ}C$$

$$\text{temperature of external air} = 3.955 - 1.958$$
$$= 1.997\,^{\circ}C$$

(The figure $1.997\,^{\circ}C$ instead of $2\,^{\circ}C$ is due to rounding off the previous values.)

Estimating condensation risk

In example 2.2 the dewpoint temperature was assumed, but The Building Research Establishment *Digest* 110 (Condensation) has made recommendations to form a basis for design which takes into account moisture content and ventilation rates.

At normal ventilation rates, the gain by the indoor air of body moisture from persons not engaged in physical exertion is roughly 4.5 g per person per hour. This results in the indoor air having an excess moisture content over outdoor air of 1.7 g of water vapour per kg of dry air.

Provided that ventilation rates are properly controlled, there would be a suitable design assumption for shops, offices, classrooms, assembly halls and dry industrial premises.

For dwellings, taking into account the moisture produced by cooking, etc., and the possible restricted ventilation in cold weather, a safer design value for excess moisture may be 3.4 g/kg.

Catering establishments and industrial workshops requiring humid atmospheres may contribute 6.8 g/kg or more to the inside air. In naturally ventilated premises, such design values may be added to assumed mixing ratio of the outdoor air.

The principles used in the *Digest* are to predict the likelihood of condensation, and to design so as to avoid this it may be applied to walls, floors, or roofs, but lightweight sheeted roofs present special problems.

Example 2.3 (see Fig. 2.3). *A cavity wall consists of 105 mm external brick leaf, 100 mm aerated concrete inner leaf, 16 mm of plaster and a 50 mm unventilated cavity. The internal and external air temperatures are 20°C and 0°C respectively. Calculate the U value and plot the structural temperature and dewpoint temperature through the wall. Check if interstitial condensation may take place. Use the following values:*

Thermal conductivities

Brick	*1.20 W/m$^{\circ}$C*
Aerated concrete	*0.14 W/m$^{\circ}$C*
Plaster	*0.40 W/m$^{\circ}$C*

Surface resistances

R_{si}	*Internal surface layer*	*0.123 m^2 $^{\circ}$C/W*
R_{so}	*External surface layer*	*0.053 m^2 $^{\circ}$C/W*
R_a	*Air space*	*0.18 m^2 $^{\circ}$C/W*

Outside air saturated at a mixing ratio of 3.8 g/kg

1. The U value

$$U = \frac{1}{R_{si} + R_{so} + L_1/k_1 + L_2/k_2 + L_3/k_3 + R_a}$$

$$U = \frac{1}{0.123 + 0.053 + 0.105/1.20 + 0.100/0.140 + 0.016/0.40 + 0.18}$$

$$U = \frac{1}{0.123 + 0.053 + 0.087 + 0.714 + 0.04 + 0.18}$$

$$U = \frac{1}{1.1975}$$

$$= 0.835\ W/m^2\,^{\circ}C$$

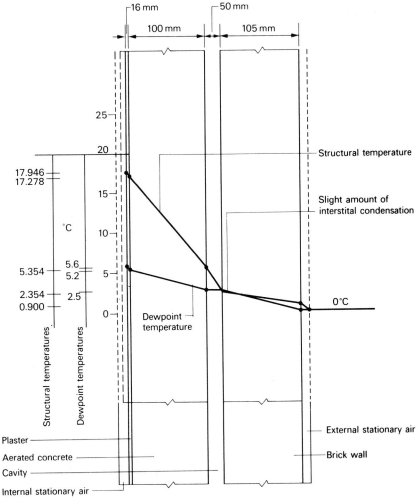

Fig. 2.3 Graph of structural and dewpoint temperatures

2. Rate of heat loss through unit area

Heat loss $= U \times$ temperature difference
$$= 0.835 \times 20$$
$$= 16.7 \text{ watts}$$

3. Thermal resistances of structural components

    (a) Internal stationary air   0.123 m^2 °C/W
    (b) Plaster   0.040 m^2 °C/W
    (c) Aerated concrete   0.714 m^2 °C/W
    (d) Brick   0.087 m^2 °C/W
    (e) External stationary air   0.053 m^2 °C/W
    (f) Air space   0.180 m^2 °C/W

4. Calculation of temperatures

Since the heat loss of 16.7 watts must pass through each part of the cavity wall, taking each part separately, its thermal transmission (U) must be equal to 1/ resistance. The temperature at various points of the wall may therefore be found as follows:

$$\text{rate of heat loss } = U \times \text{temperature difference}$$

also $\text{rate of heat loss } = \dfrac{1}{R} \times \text{temperature difference}$

5. Internal stationary air or plaster surface

$$16.7 = \frac{1}{0.123} \times \text{temperature difference}$$

temperature difference $= 16.7 \times 0.123$
$$= 2.054 \,°\text{C}$$
temperature of plaster surface $= 20 - 2.054$
$$= 17.946 \,°\text{C}$$

6. Inner face of aerated concrete leaf

$$16.7 = \frac{1}{0.04} \times \text{temperature difference}$$

temperature difference $= 16.7 \times 0.04$
$$= 0.668 \,°\text{C}$$
temperature at inner face $= 17.946 - 0.668$
$$= 17.278 \,°\text{C}$$

7. Opposite face of aerated concrete leaf

$$16.7 = \frac{1}{0.714} \times \text{temperature difference}$$

temperature difference $= 16.7 \times 0.714$
$$= 11.924 \,°\text{C}$$
temperature at opposite face $= 17.278 - 11.924$
$$= 5.354 \,°\text{C}$$

8. Inner face of brick leaf

$$16.7 = \frac{1}{0.18} \times \text{temperature difference}$$

temperature difference $= 16.7 \times 0.18$
$$= 3 \,°\text{C}$$
temperature at inner face $= 5.354 - 3$
$$= 2.354 \,°\text{C}$$

9. Outer face of brick leaf

$$16.7 = \frac{1}{0.087} \times \text{temperature difference}$$

temperature difference $= 16.7 \times 0.087$
$$= 1.453 \,°\text{C}$$

temperature at outer face = 2.354 − 1.453
                         = 0.9 °C

10. External stationary air

$$16.7 = \frac{1}{0.053 \times \text{temperature difference}}$$

temperature difference = 16.7 × 0.053
                       = 0.885 °C
temperature outside = 0.9 − 0.885
                    = 0.015 °C

(The difference between the temperature given in the question as 0 °C is due to rounding off the figures.)

The section through the cavity wall can now be drawn to a suitable scale and the temperature difference across the structure set up on an adjacent vertical scale so that the various temperatures may be plotted.

Dewpoint temperature

A psychrometric chart is used to estimate dewpoints (see Fig. 2.4). The chart enables the effect of temperature changes and dewpoint temperatures across the wall to be predicted.

Condensation risk

At any point where the estimated structural temperature is lower than the dew-point temperature, condensation may occur. In order to find the vapour-pressure drop and the corresponding dewpoint temperatures between points of the structure, the vapour resistivity of the various materials must be known. Table 2.1 gives both thermal and vapour resistivity of some common materials.

Table 2.1

Materials	Thermal resistivity (m²C/W)	Vapour resistivity (MNs/g)
Brickwork	0.7−1.4	25−100
Concrete	0.7	30−100
Rendering	0.8	100
Plaster	2	60
Timber	7	45−75
Plywood	7	1500−6000
Fibre building board	15−19	15−60
Hardwood	7	450−750
Plasterboard	6	45−60
Compressed strawboard	10−12	45−75
Wood-wool slabs	9	15−40
Expanded polystyrene	30	100−600
Foamed urea-formaldehyde	26	20−30
Foamed polyurethane (open or closed cell)	40−50	30−1000
Expanded ebonite	34	11 000−60 000

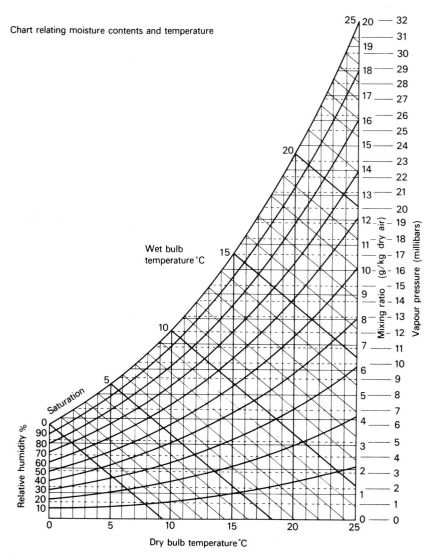

Chart relating moisture contents and temperature

Fig 2.4   Psychrometric chart

1. From the psychrometric chart (Fig. 2.4) and using the design assumption that the external air is at 0 °C and is saturated at a mixing ratio of 3.8 g/kg, the outdoor vapour pressure can be read (on the right-hand scale) as 6 mb. If a moisture vapour excess of 1.7 g/kg is contributed to this by activities indoors, the two right-hand scales show that for a total moisture content of 3.8 + 1.7 = 5.5 g/kg and the indoor vapour pressure will be 9 mb. The difference between the indoor and outdoor vapour pressure is 9 − 6 = 3 mb, therefore ΔP = 3 mb.

2. Vapour resistance
   (Value from Table 2.1 × thickness of component)

Plaster	60×0.016	$= 0.96$
Aerated concrete	30×0.100	$= 3.00$
Cavity	Nil	$= 0.00$
Brick	25×0.105	$= 2.625$
	R_v	$= 6.585$

3. Pressure drop between points

$$\frac{\Delta P}{R_v} \times r_v = \frac{3}{6.93} \times r_v = 0.133\, r_v \text{ (mb point to point)}$$

R_v = Total vapour resistance

r_v = Individual vapour resistances.

			Corresponding dewpoint temperature from chart ($^\circ$C)
Indoor vapour pressure		9.000	$= 5.6\,^\circ$C
across plaster 0.433×0.96	$= 0.416$		
		8.584	$= 5.2\,^\circ$C
drop to			
across inner leaf 0.433×3	$= 1.299$		
drop to		7.285	$= 2.5\,^\circ$C
across cavity Nil	$= 0.000$		
		7.285	$= 2.5\,^\circ$C
across outer leaf 0.433×2.625	$= 1.137$		
drop to		6.148	$= 0.0\,^\circ$C

Figure 2.3 shows a graph of these structural and dewpoint temperatures.

Note: If the outer portion of the wall is permeable to moisture, or if ventilation is provided behind permeable wall or roof claddings, condensation will not be troublesome because the moisture can evaporate gradually to the outside air. If the outer surface is impermeable, the condensed moisture tends to accumulate in the wall and may ultimately saturate the material. The situation will be most severe when the relative humidity inside is high.

Vapour barrier

A vapour barrier on the inner face of the wall (or the potentially warm side of any layer of insulating material) will prevent the passage of water vapour into the wall. If the outer face of the wall is more permeable than the inner vapour barrier, any moisture contained in the wall can evaporate to the outside air.

If, however, the outer face of the wall has an impermeable cladding, or if the cladding is of organic material (for example, timber) that would suffer in prolonged damp conditions, a ventilated cavity should be formed between the cladding and the wall so that any moisture evaporating from the wall surface is removed.

Surface temperature

The temperature at any point within a structure is proportional to the ratio

which the thermal resistance at that point bears to the total thermal resistance of the structure.

If the U value is not required, the thermal resistance may be used to find the temperature drop through the structure:

$$\frac{R_s}{R_T} = \frac{t_s}{t_T}$$

where R_s = surface resistance

R_T = total thermal resistance of structure

t_s = temperature difference between the surface and the air

t_T = total temperature difference between inside and outside air

---

Example 2.4. *Find the internal temperature of 4 mm single glazing when the external temperature is $2\,^\circ$C and the internal temperature is $20\,^\circ$C.*
Use the following data:

external surface resistance	*$0.053\ m^2\,^\circ C/W$*
internal surface resistance	*$0.123\ m^2\,^\circ C/W$*
thermal conductivity of glass	*$1.02\ W/m\,^\circ C$*

First find the total thermal resistance.

external surface resistance	0.053
internal surface resistance	0.123
thermal resistance of glass	
$= \dfrac{L}{k} = \dfrac{0.004}{1.02} =$	0.004
total thermal resistance	0.180

$$\frac{R_s}{R_T} = \frac{t_s}{t_T}$$

$$\frac{0.123}{0.180} = \frac{t_s}{(20-2)}$$

$$\therefore \quad 0.683 = \frac{t_s}{18}$$

$$t_s = 0.683 \times 18$$
$$= 12.294\,^\circ C$$

internal surface
temperature $= 20 - 12.3$
$= 7.7\,^\circ C$ (approx.)

Example 2.5. *If in example 2.4 the wet-bulb temperature of the inside air is $14\,^\circ$C, find by use of the psychrometric chart if condensation will occur on the inside surface of the glass.*

By intersecting the dry-bulb and wet-bulb temperature lines as shown in Fig. 2.4, it will be seen that the relative humidity of the inside air is about 50 per cent.

If a horizontal line is now drawn across from the point of intersection of the dry-bulb and wet-bulb temperature lines until it intersects the saturation curve, it will be seen that the temperature at which condensation will occur is about 9.8 °C. Condensation will therefore occur on the inside surface of the glass.

Example 2.6. *If the window in example 2.4 is double-glazed by adding another 4 mm thick pane of glass and leaving a 20 mm sealed air space between the two panes, find the surface temperature of the inner pane of glass.*

Use the same values given in example 2.4 but include:

thermal conductivity of air = 0.029 W/m °C

If the temperature of the inside air is to be 20 °C with a relative humidity of 50 per cent check by use of the psychrometric chart if condensation on the inside surface of the glass has now been prevented.

Total thermal resistance:

external surface resistance	= 0.053
internal surface resistance	= 0.123

thermal resistance of glass

$$= \frac{L}{k} = \frac{0.008}{1.02} \qquad = 0.008$$

thermal resistance of air space

$$= \frac{L}{k} = \frac{0.02}{0.029} \qquad = 0.689$$

total thermal resistance $\qquad = \overline{0.873}$

$$\frac{R_s}{R_T} = \frac{t_s}{t_T}$$

$$\frac{0.123}{0.873} = \frac{t_s}{(20 - 2)}$$

$$t_s = 0.141 \times 18$$

$$= 2.538 \,°C$$

internal surface temperature $= 20 - 2.538$

$$= 17.5 \,°C \text{ (approx.)}$$

Since the air temperature is higher than the dewpoint temperature of 9.8 °C, condensation should not now occur.

Example 2.7. *A roof is to be constructed of 10 mm thick asbestos-cement sheets fixed to a steel frame and insulated with 50 mm thick fibre board to form an air space. If the inside and outside air temperatures are 20 °C and 0 °C respectively, calculate the U value and the inside surface temperature. Check by use of the psychrometric chart if condensation will occur on the inside surface of the insulation when the relative humidity of the inside air is 60 per cent.*

Use the following data:

thermal resistance of external air	*0.045*	*m² °C/W*
thermal resistance of internal air	*0.110*	*m² °C/W*

thermal resistance of air space	*0.18*	*m² °C/W*
thermal resistivity of fibre board	*19*	*m °C/W*
thermal resistivity of asbestos cement	*4.5*	*m °C/W*

$$U = \frac{1}{R_{ao} + R_{ai} + R_a + (L_1 \times r_1) + (L_2 \times r_2)}$$

$$U = \frac{1}{0.045 + 0.110 + 0.18 + (0.01 \times 4.5) + (0.05 \times 19)}$$

$$U = \frac{1}{0.045 + 0.110 + 0.18 + 0.045 + 0.95}$$

$$U = \frac{1}{1.33}$$

$$= 0.752 \text{ W/m}^2 \,°C$$

rate of heat loss $= 0.752 \times 20$

$$= 15.04 \text{ W/m}^2$$

Temperature of inside surface

rate of heat loss $= \frac{1}{R} \times$ temperature difference

temperature difference $= 15.04 \times 0.110$

$$= 1.654 \,°C$$

temperature of inside surface $= 20 - 1.654$

$$= 18.35 \,°C$$

Inspection of the psychrometric chart (Fig. 2.4) shows that for 20 °C indoor air temperature and 60 per cent relative humidity, the dewpoint temperature of the air is about 11.75 °C. The structural temperature of 18.35 °C is above the dewpoint temperature of 11.75 °C and therefore condensation on the surface is unlikely to occur. A vapour barrier behind the fibre-board insulation would, however, be a recommendation in case of interstitial condensation.

Alternatively, the temperature of the inside surface may be found from the thermal resistances as follows:

$$\frac{R_s}{R_T} = \frac{t_s}{t_T}$$

$$\frac{0.11}{1.33} = \frac{t_s}{20}$$

$$0.0827 = \frac{t_s}{20}$$

$$\therefore \qquad t_s = 0.827 \times 20$$

$$= 1.654 \,°C$$

temperature of inside surface $= 20 - 1.654$

$$= 18.35 \,°C$$

Example 2.8. *A factory is to be constructed of corrugated sheeting fixed to a steel frame and insulated with fibre board lining to form an air space. It is required to provide sufficient thickness of insulation to prevent the inside* *surface from falling below 15 °C when the inside and outside air temperatures are 20 °C and 2 °C respectively.*

Using the following data, determine the minimum thickness of the insulation and the overall thermal transmittance.

U values

thermal transmittance of sheeting	$8 \ W/m^2 \ °C$
thermal resistance of air space	$0.180 \ m^2 \ °C/W$
thermal resistance of internal surface	$0.123 \ m^2 \ °C/W$
thermal resistance of external surface	$0.053 \ m^2 \ °C/W$
thermal resistivity of fibre board	$19 \ m \ °C/W$

Thermal resistance of sheeting

$$\text{rate of heat loss} = U \times \text{temperature difference}$$
$$= 8 \times [20 - (-2)]$$
$$= 8 \times 22$$
$$= 176 \ W/m^2$$

$$\text{rate of heat loss} = \frac{1}{R} \times \text{temperature difference}$$

$$176 = \frac{1}{R} \times 22$$

$$\frac{176}{22} = \frac{1}{R}$$

$$\therefore \quad R = \frac{2}{176}$$

$$= 0.125 \ m^2 \ °C/W$$

thermal resistance of fibre board = resistivity × thickness in metres

$$\text{total resistance } R_T = 0.123 + 0.053 + 0.125 + 0.18 + (r \times \text{thickness})$$
$$= 0.481 + (r \times \text{thickness})$$
$$= 0.481 + (19 \times \text{thickness})$$

$$\frac{R_s}{R_T} = \frac{t_s}{t_T}$$

$$\frac{0.123}{0.481 + (19 \times \text{thickness})} = \frac{15}{22}$$

$$0.481 + (19 \times \text{thickness}) \times 15 = 0.123 \times 22$$

$$0.481 + (19 \times \text{thickness}) = \frac{0.123 \times 22}{15}$$

$$19 \times \text{thickness} = \frac{0.123 \times 22}{15} - 0.481$$

$$\text{thickness} = \frac{0.1804 - 0.481}{19}$$

$$= 0.016 \ m$$

$$= 16 \ mm$$

$$U = \frac{1}{0.123 + 0.053 + 0.125 + 0.18 + (19 \times 0.016)}$$

$$U = \frac{1}{0.123 + 0.053 + 0.125 + 0.18 + 1.14}$$

$$U = 1.621 \ W/m^2 \ °C$$

Questions

1. Find the internal and external surface temperatures of a 114 mm thick solid brick wall when the internal and external air temperatures are 20 °C and 2 °C respectively. Use the following data:

external surface resistance	$0.053 \ m^2 \ °C/W$
internal surface resistance	$0.123 \ m^2 \ °C/W$
thermal conductivity of brick	$1.20 \ W/m \ °C$

   Answers: internal surface of brick wall = 11.83 °C; external surface of brick wall = 5.52 °C

2. Define the following forms of condensation: (*a*) temporary; (*b*) permanent; (*c*) interstitial.

3. Find the internal and external surface temperatures of a cavity wall having a 105 mm thick brick outer leaf, 50 mm cavity and a 38 mm thick mineral wool slab inner leaf. Use the following values:

external surface resistance	$0.053 \ m^2 \ °C/W$
internal surface resistance	$0.123 \ m^2 \ °C/W$
thermal conductivity of brick	$1.20 \ W/m \ °C$
thermal resistance of air space	$0.18 \ m^2 \ °C/W$
thermal resistivity of mineral wool	$8.66 \ m \ °C/W$
internal and external air temperatures 20 °C and 0 °C respectively	

   Answers: (*a*) internal surface = 16.8 °C
   (*b*) opposite side of insulation = 8.27 °C
   (*c*) inner surface of brick leaf = 3.62 °C
   (*d*) outer surface of brick leaf = 1.37 °C

4. Find the internal temperature of 8 mm thick single glazing when the internal and external air temperatures are 20 °C and −1 °C respectively. Use the following data:

external surface resistance	$0.053 \ m^2 \ °C/W$
internal surface resistance	$0.123 \ m^2 \ °C/W$
thermal conductivity of glass	$1.20 \ W/m \ °C$

   Answer: 5.7 °C (approx.)

5. If the window in question 4 is double-glazed by adding another 8 mm thick pane of glass and leaving a 20 mm sealed air space between the two panes, find the surface temperature of the inner pane of glass. Use the same values given in question 4 but including: thermal conductivity of air = 0.029 W/m °C. If the room has a relative humidity of 50 per cent determine from the

pschyrometric chart whether condensation will be deposited on the inside surface.

Answer: 17°C. There will be no condensation as the dewpoint temperature is 9.5°C.

6. A cavity wall consists of 105 mm thick brick outer leaf, 100 mm thick aerated concrete inner leaf, 16 mm internal plastering and a 50 mm cavity filled with foamed polyurethane. If the internal and external air temperatures are 22°C and 0°C respectively, find the structural and dewpoint temperatures through the wall and check the possibility of interstitial condensation taking place. Draw a section through the wall showing a graph of the temperatures. Use the following values:

thermal conductivities	
brick	1.20 W/m°C
aerated concrete	0.14 W/m°C
plaster	0.40 W/m°C
thermal resistivity of foamed polyurethane	40 m°C/W
internal surface resistance	0.123 m²°C/W
external surface resistance	0.053 m²°C/W
vapour resistances	
brick	60 MNs/g
polyurethane foam	40 MNs/g
aerated concrete	30 MNs/g
plaster	60 MNs/g

Outside air relative humidity 100 per cent. Inside activities assumed to contribute a moisture vapour excess of 3.2 g/kg.

Answer:

Position	Structural temperature (°C)	Dewpoint temperature (°C)
Surface of plaster	21.107	8.75
Inner face of aerated concrete leaf	20.817	8.00
Opposite face of aerated concrete leaf	15.633	6.20
Inner face of brick leaf	1.113	5.00
External face of brick leaf	0.483	0.00

Figure 2.5 shows a graph of structural and dewpoint temperatures.

7. A wall is to be constructed of 10 mm thick asbestos-cement sheets fixed to a steel frame and insulated with 50 mm thick fibre board lining to form an air space. If the inside and outside air temperatures are 20°C and 0°C respectively, calculate the inside surface temperature and check by use of the psychrometric chart if condensation will occur on the inside surface of the insulation when the relative humidity of the inside air is 50 per cent. Use the following data:

thermal resistance of external air	0.053 m²°C/W
thermal resistance of internal air	0.123 m²°C/W

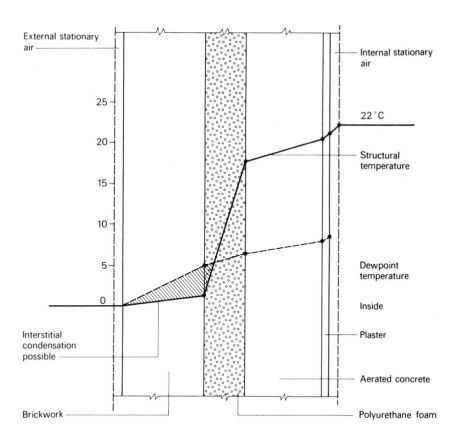

Fig. 2.5 Graph of structural and dewpoint temperatures

thermal resistance of air space	0.180 m²°C/W
thermal resistivity of asbestos cement	4.5 m°C/W
thermal resistivity of fibre board	19 m°C/W

Answers: Inside surface temperature = 18.18°C
Dewpoint temperature of inside air = 9.8°C
Condensation will therefore not occur

33

Chapter 3

Low-pressure hot-water heating

Systems

When low-pressure hot water is used for heating systems the temperature of the water is below boiling point, usually about 80 °C on the flow pipe and between 60 °C and 70 °C on the return pipe.

Water has the high specific heat capacity of 4.2 kJ/kg °C, and although it is more difficult to heat than other fuels, more heat can be transferred from the boiler to the various heat emitters with pipes of a relatively small diameter. The higher the temperature of the water the greater the amount of heat transferred, but some of this heat is lost by the higher heat losses on the pipework. This higher temperature may also cause injury to persons coming into contact with the radiators, etc., and special precautions must be taken (see Chapter 7, Medium and high-pressure hot-water systems).

Water circulation

Water may be circulated in the system either by natural convection (thermo-siphonage) or by means of a centrifugal pump. Thermo-siphonage is produced by the difference in temperature between the flow and return pipes. The denser cooler water in the return pipe forces the less dense hotter water in the flow pipe through the circuit. Pump circulation has now replaced thermo-siphonage circulation for all but the smallest house installation. It has the advantage of reducing the heating-up period and also smaller pipes may be used.

Circuit arrangement

Various circuits may be used depending upon the type and layout of the building. In some buildings two or more circuits may be used supplied from the same boiler plant. One-pipe or two-pipe circuits may be used and although the one-pipe is simpler than the two-pipe it has two disadvantages:

1. The cooler water passing out of each heat emitter flows to the next one, resulting in the emitters at the end of the circuit being cooler than those at the beginning. This can be reduced by shutting down the lock shield valves at the start of the circuit thus allowing more water to pass to the emitters at the end of the circuit.
2. Even if the circuit is pumped, the pump pressure does not force water through the heat emitters. Water is induced to circulate through the emitters by thermo-siphonage created by the difference in density between the water entering and leaving the emitters. A one-pipe system therefore cannot be used to supply heat to various types of emitters that have a comparatively high resistance to the flow of water.

Injector tees can be used to induce water through emitters on a one-pipe circuit, but these increase the resistance to the flow of water.

The two-pipe system requires more pipework, but this can be reduced in diameter as it passes around the circuit. The pump pressure acts throughout the circuit and therefore any type of heat emitter can be used. There is much less cooling down of the emitters at the end of the circuit, due to the cooler water passing out of the emitters being returned to the boiler via the return pipe.

One-pipe ring circuit (Fig. 3.1): suitable for small single-storey buildings and if the circuit is pumped the heat emitters may be on the same level as the boiler. All the other circuits shown may be used for multi-storey buildings.

One-pipe ladder (Fig. 3.2): suitable when it is possible to use horizontal runs either exposed in a room or in a floor duct.

One-pipe drop (Fig. 3.3): this circuit requires provision at the top of the building, for a horizontal main distributor pipe and at the bottom of the building, for a main horizontal collector pipe. It is suitable for offices, etc., having separate rooms on each floor where the vertical pipe can be housed in a corner duct. It has the advantage of making the heat emitters self venting.

One-pipe parallel (Fig. 3.4): similar to the ladder system and is used when it is impractical to install a vertical main return pipe.

Two-pipe parallel (Fig. 3.5): again similar to the ladder system, but possessing the advantages of a two-pipe circuit.

Two-pipe upfeed (Fig. 3.6): suitable for buildings having floors of various heights, or for groups of buildings. An automatic vent valve can be used at the top of each flow pipe. The system is also used for embedded panel heating, using a panel of pipes instead of radiators.

Two-pipe high level return (Fig. 3.7): used when it is impractical to install a main horizontal collector pipe. The system is most useful when installing heating in existing buildings having a solid ground floor.

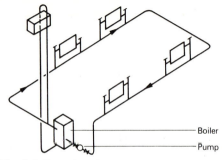

Fig. 3.1 One-pipe ring

- Boiler
- Pump

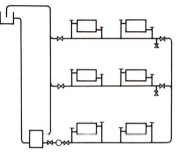

Fig. 3.2 One-pipe ladder

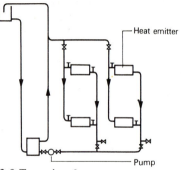

- Heat emitter
- Pump

Fig. 3.9 Two-pipe drop

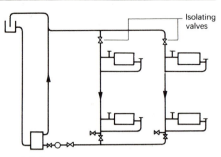

- Isolating valves

Fig. 3.3 One-pipe drop

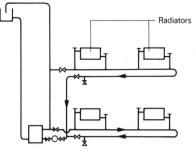

- Radiators

Fig. 3.4 One-pipe parallel

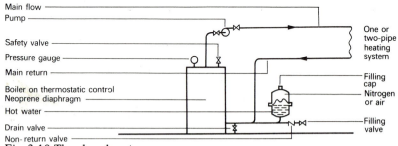

- Main flow
- Pump
- Safety valve
- Pressure gauge
- Main return
- Boiler on thermostatic control
- Neoprene diaphragm
- Hot water
- Drain valve
- Non-return valve
- One or two-pipe heating system
- Filling cap
- Nitrogen or air
- Filling valve

Fig. 3.10 The closed system

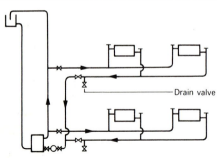

Fig. 3.5 Two-pipe parallel

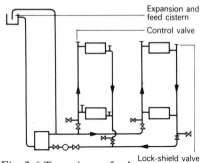

- Expansion and feed cistern
- Control valve
- Lock-shield valve

Fig. 3.6 Two-pipe upfeed

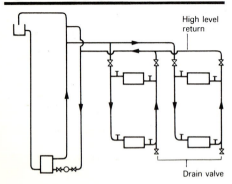

- High level return
- Drain valve

Fig. 3.7 Two-pipe high-level return

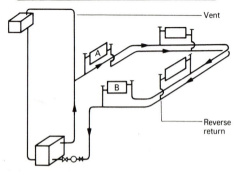

- Vent
- A
- B
- Reverse return

Fig. 3.8 Two-pipe reverse return

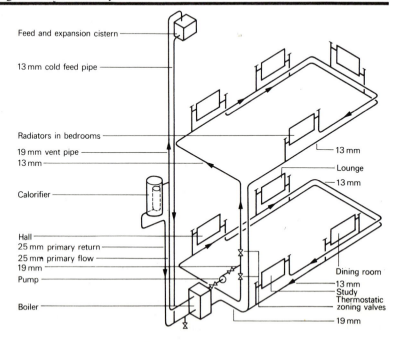

- Feed and expansion cistern
- 13 mm cold feed pipe
- Radiators in bedrooms
- 19 mm vent pipe
- 13 mm
- Calorifier
- Hall
- 25 mm primary return
- 25 mm primary flow
- 19 mm
- Pump
- Boiler
- 13 mm
- Lounge
- 13 mm
- Dining room
- 13 mm Study
- Thermostatic zoning valves
- 19 mm

Fig. 3.11 Small-bore heating system

Two-pipe reverse return (Fig. 3.8): the predominant feature of this system is the equal travel to each heat emitter which provides a well-balanced circuit. Note that the length of the circuit for emitter A is the same as for emitter B.

Two-pipe drop (Fig. 3.9): suitable for buildings where it is possible to install a main horizontal distributor pipe and a main horizontal collector pipe in ducts. The system is self venting and therefore periodic venting of the radiators is unnecessary.

Isolating valves

Each heat emitter should be provided with a control valve on the inlet and a lock shield valve on the outlet. The lock shield valve is used by the heating engineer to balance the circuit as described earlier.

Isolating valves will also be required on the boiler, pumps and on all main branches. Drain valves must also be provided, to allow the various sections to be drained down.

Expansion and feed cistern

When the system is cold the ball valve should be set so that there is not more than 100 mm of water in the cistern and the capacity of the cistern should be such that when the water is raised to its working temperature the water level does not rise within 50 mm from the bottom of the warning pipe.

Feed and vent pipes

The feed pipe should be of adequate diameter and taken from the bottom of the expansion cistern, without any branch pipes, direct to the bottom of the boiler or return pipe. If a valve is fitted to the feed pipe, it should be of the lock shield type to prevent unauthorised use.

The vent pipe should be taken from the boiler and turned over the expansion cistern. A valve should never be fitted on the vent pipe, unless a three-way type is used when two or more boilers are to be installed (see centralised hot water supply).

Position of pump

This may be fitted on either the main flow or main return pipe. If fitted on the flow pipe, the heating system is under positive pressure and there is less risk of air being drawn into the circuit.

If fitted on the return pipe, the pump is normally easier to install and is at a lower temperature. There is however a greater risk of drawing air into the circuit, due to negative pump pressure.

Closed systems (Fig. 3.10)

In place of an expansion and feed cistern, the expansion of the heated water may be accommodated in a closed expansion vessel fitted to the boiler. The vessel contains nitrogen or air above a neoprene diaphragm and as the water expands the gas is compressed and its pressure rises. The vessel should accommodate the expansion of the heated water in the system.

Small-bore heating systems (Fig. 3.11)

The system is used extensively for houses and other types of small buildings. Use is made of small-bore copper pipes, usually 13 mm or 19 mm bore, depending upon the heating load to be carried in the circuit.

The boiler also heats the hot-water calorifier by natural convection. Thermostatic control may be by zoning valves shown, separate thermostatic radiator valves, a three-way mixing valve or by switching on and off the pump by a wall thermostat.

Thermal comfort, hot water and steam heat emitters

Thermal comfort

The thermal comfort of a human body is governed by the following factors:

1. The heat lost in radiation from the body, through clothing and exposed skins surfaces to the cooler surroundings.
2. The heat lost by convection from the body through clothing and exposed skin surfaces due to contact with the surrounding air, the temperature of which is considerably lower than that of the body.
3. The heat lost from the body by evaporation from the skin, due to perspiring:

The normal losses from the body, from the above sources, approximate to the following:

Radiation	45 per cent
Convection	30 per cent
Evaporation	25 per cent

In order to preserve the normal temperature of the body, these heat losses must be balanced by the heat gained. In the absence of sufficient warmth from the sun, or heat gains inside the building from lighting, people or machines, the necessary heat must be provided by heat emitters. It follows therefore that for various types of human activity there must be the correct proportions of radiant and convection heat, which will provide the most comfortable artificial warmth.

The rate of heat losses from the body can be controlled by the following:

1. *Radiation* — by the mean radiant temperature of the surrounding surfaces.
2. *Convection* — by the air temperature and air velocity.
3. *Evaporation* — by the relative humidity of the air and air velocity.

The purpose of a heat emitter is to maintain, at economic cost, the conditions of mean radiant temperature, air temperature and velocity that will give a suitable balance between the three ways in which heat is lost from the body. Although a heat emitter will provide radiant and convection heating, the correct control of relative humidity may require a system of air conditioning, which is described on p. 54.

Figure 3.12 shows how the heat losses and heat gains from and to a person in a room are balanced, also the external heat gains and losses.

Hot water and steam heat emitters

A centralised hot-water heating system has three basic elements:

1. Boiler plant for heat generation.
2. Heat distribution circuit.
3. Heat emitter.

The main type of heat emitters used for centralised hot water heating systems are:

1. Radiators

These may be column, hospital or panel types, made from either steel or cast iron. Steel radiators are made from light gauge steel pressings welded together, they are modern in appearance and are used extensively for heating systems in houses and flats. Cast iron radiators are bulkier and heavier, but will stand up to rough use in schools, hospitals and factories.

If a radiator is fitted against a wall, staining of the wall above the radiator will occur due to convection currents picking up dust from the floor. To prevent this, a shelf should be fitted about 76 mm above the radiator, and the jointing of the shelf to the wall must be well made or otherwise black stains will appear above the shelf. End shields must be fitted to the shelf to prevent black stains at the sides.

Painting of radiators: The use of metallic paints reduce the heat emitted from a radiator and the best radiating surface is dead black. Any colour of non-metallic paints may be used, as these do not affect appreciably the amount of radiant heat emitted by the radiator. The heat emitted by convection is not affected by the painting of radiators. The name 'radiator' is misleading, for although heat is transmitted by radiation, a greater proportion of heat is transmitted by convection, depending upon the type of radiator used.

Position of radiators: The best position is under a window, so that the heat emitted mixes with the incoming cold air from the window and this prevents cold air passing along the floor, which would cause discomfort to the occupants of the room. Figures 3.13, 3.14 and 3.15 show column, hospital and panel types of radiators. Figure 3.16 shows a radiator shelf.

2. Radiant panels

These have flat faced metal fronts and are similar to panel radiators, but transmit a greater proportion of heat by radiation. The panels are particularly suited to the heating of workshops, where they may be suspended at heights from 3 to 4 m above the floor level and arranged so that the heat is radiated downwards. They have the advantage of giving comfortable conditions to the occupants of the room by providing radiant heat at a lower air temperature. There is also a lower temperature gradient between the floor and the ceiling, which added to the lower air temperature reduces the cost of heating by about 15 per cent.

Figure 3.17 shows a radiant panel made from steel and Fig. 3.18 shows the various positions of the panels in a workshop.

3. Natural convection

These can be cabinet or skirting types: the cabinet type comprises a finned tubular heating element fitted near to the bottom of the casing, so that a stack effect is created inside the cabinet. The column of warm air above the heating element is displaced through the top of the cabinet by the cool air entering at the bottom. The greater the height of the cabinet the greater will be the air flow through it. Skirting types provide a good distribution of heat in a room and are very neat in appearance. If the floor is to be carpeted care should be taken to ensure that a gap is left at the bottom of the heater casing or convection through the heater will be prevented. Figure 3.19 shows a cabinet type convector and Fig. 3.20 shows a skirting type convector.

4. Fan convectors

These have a finned tubular heating element, usually fitted near the top of the casing. The fan or fans fitted below the element draw air in from the bottom of the casing, the air is then forced through the heating element, where it is heated, before being discharged through the top of the cabinet into the room. The fans may have two- or three-speed control and thermostatic control of the heat output is often by switching off, or a change of speed of the fans. If required, a clock can also control the switching on and off of the fans. The convector may be fitted with an air filter above the heating element, which is not possible with the natural convector. They also have the advantage of quickly heating the air in the room and give a good distribution of heat. Figure 3.21 shows a fan convector incorporating an air filter.

5. Overhead unit heaters

These are similar to the fan convector and have a finned tubular heating element with a fan to improve the circulation of warm air. The method of fixing the fan and the heating element however differs from the fan convector to permit the overhead installation of the heater in factories, garages and warehouses, so that the warm air is blown down on to the working area. They usually operate on high temperature hot water or steam and the louvres can be adjusted so as to alter the direction of the warm air. Figure 3.22 shows an overhead unit heater and Fig. 3.23 shows how several heaters may be used to heat a workshop.

Note: The heating efficiency of both types of convectors and unit heaters is lowered more quickly than that of radiators, due to a drop in temperature of the heating element. For this reason, it is better to use a two-piped pumped distribution system.

6. Overhead radiant strips

These overcome the difficulty and cost of connecting separate radiant panels at high level. They consist of heating pipes, up to 30 m long, fixed to an insulated metal plate which also becomes heated by conduction from the pipe. The minimum mounting height of the strips is governed by the heating system temperature and ranges from about 3 m for low temperature hot water, to about 5 m for high temperature hot water and steam. Figure 3.24 shows a radiant strip having two heating pipes; one-to-four heating pipes may also be used. Figure 3.25 shows how radiant strips may be installed in a workshop.

7. Embedded pipe panels

Continuous coils of copper or steel pipes of 13 or 19 mm bore at 225 to 300 mm centres are embedded inside the building fabric, usually in the floor or ceiling, although wall panels may also be used. The following recommend panel surface temperatures may be used, except in special cases, such as may be required for entrance halls to public buildings or areas bordering on exposed outside walls.

Floors	26.7 °C
Ceilings	49 °C
Walls	43 °C

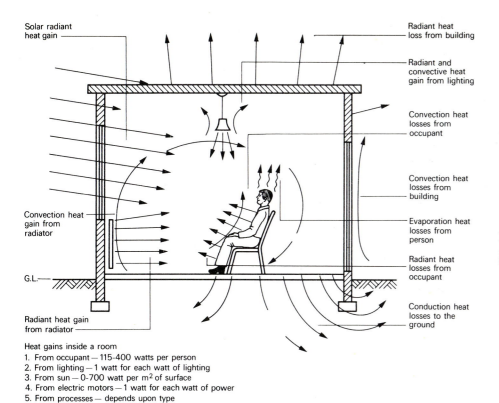

Solar radiant heat gain

Radiant heat loss from building

Radiant and convective heat gain from lighting

Convection heat losses from occupant

Convection heat losses from building

Evaporation heat losses from person

Radiant heat losses from occupant

Conduction heat losses to the ground

Convection heat gain from radiator

G.L.

Radiant heat gain from radiator

Heat gains inside a room
1. From occupant — 115-400 watts per person
2. From lighting — 1 watt for each watt of lighting
3. From sun — 0-700 watt per m² of surface
4. From electric motors — 1 watt for each watt of power
5. From processes — depends upon type

Fig. 3.12 Heat balance for an occupant in a room, and the external heat gains and losses

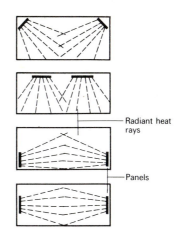

Radiant heat rays

Panels

Fig. 3.18 Positions of radiant panels

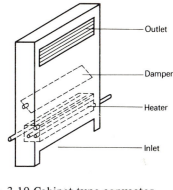

Outlet

Damper

Heater

Inlet

Fig. 3.19 Cabinet type convector

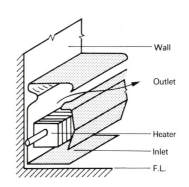

Wall

Outlet

Heater

Inlet

F.L.

Fig. 3.20 Skirting type convector

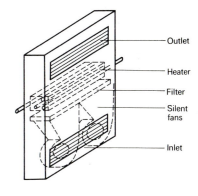

Outlet

Heater

Filter

Silent fans

Inlet

Fig. 3.21 Fan convector

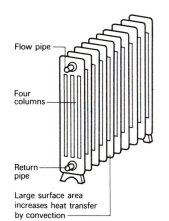

Flow pipe

Four columns

Return pipe

Large surface area increases heat transfer by convection

Fig. 3.13 Column type radiator

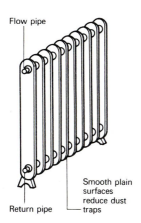

Flow pipe

Return pipe

Smooth plain surfaces reduce dust traps

Fig. 3.14 Hospital pattern radiator

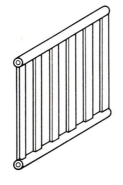

Fig. 3.15 Panel radiator made from pressed steel

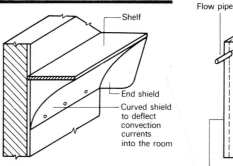

Shelf

End shield

Curved shield to deflect convection currents into the room

Fig. 3.16 Radiator shelf

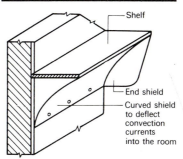

Insulation at rear

Flow pipe

Return pipe

Pipe coil

Flat metal front

Flat metal rear and side

Fig. 3.17 Radiant panel

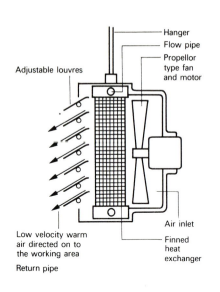

Hanger

Flow pipe

Propellor type fan and motor

Adjustable louvres

Low velocity warm air directed on to the working area

Return pipe

Air inlet

Finned heat exchanger

Fig. 3.22 Overhead unit heater

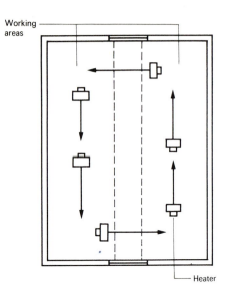

Working areas

Heater

Fig. 3.23 Arrangement of unit heaters

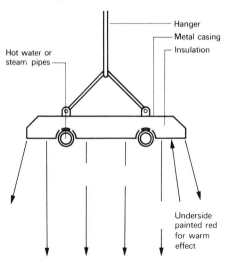

Hanger

Metal casing

Insulation

Hot water or steam pipes

Underside painted red for warm effect

Radiant heat directed onto working area

Fig. 3.24 Overhead radiant strip

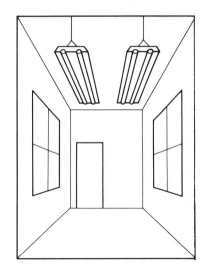

Fig. 3.25 Arrangement of radiant strips

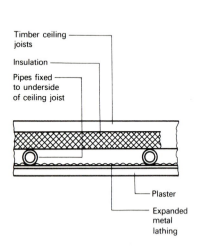

Timber ceiling joists

Insulation

Pipes fixed to underside of ceiling joist

Plaster

Expanded metal lathing

Fig. 3.26 Ceiling panel

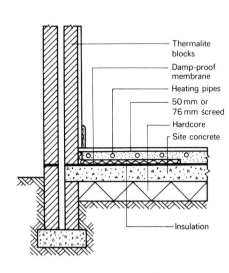

Thermalite blocks

Damp-proof membrane

Heating pipes

50 mm or 76 mm screed

Hardcore

Site concrete

Insulation

Fig. 3.27 Floor panel

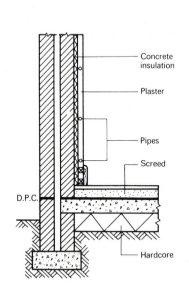

Concrete insulation

Plaster

Pipes

Screed

D.P.C.

Hardcore

Fig. 3.28 Wall panels

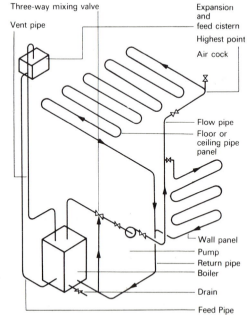

Three-way mixing valve

Vent pipe

Expansion and feed cistern

Highest point

Air cock

Flow pipe

Floor or ceiling pipe panel

Wall panel

Pump

Return pipe

Boiler

Drain

Feed Pipe

Fig. 3.29 Embedded panel system

With regards to these panel temperatures, the required water temperatures in the flow and return mains for the different panel locations are:

Panel location	Flow pipe	Return pipe
Ceilings	54 °C	43 °C
Floor	43 °C	35 °C
Wall	43 °C	35 °C

In order to control the temperature of the pipe panel, a modulating three-way mixing valve is used which allows the cooler water in the return pipe to mix with the hotter water in the flow pipe. The amount of radiant heat given off to the room, from different panel locations, is as follows:

Ceilings	65 per cent
Floors	50 per cent
Walls	50 per cent

Since radiant heat provides a feeling of warmth at a lower temperature than heat provided by convection, ceiling panels are to be preferred.

Figures 3.26, 3.27 and 3.28 show the methods of embedding the pipes in ceilings, floors and walls. Figure 3.29 shows an embedded panel system, including a three-way mixing valve.

Pipework

Copper pipes are usually jointed by means of soft soldered capillary fittings, but silver soldered or bronze welded joints are also suitable; steel pipes are welded. Compression joints for copper and screwed joints for steel are not to be recommended, due to risks of leaks caused by vibration of the building fabric. Before the pipes are embedded, they should be hydraulically tested at a pressure of 1400 kPa and this pressure should be maintained for 24 hours. The floor screed should be allowed to cure naturally before the pipes are heated and then heated gradually over a period of 10 days, before being put on full load.

Merits of heating by radiation

Radiant heating

1. Since about 45 per cent of the heat lost by the human body is due to radiation, the feeling of warmth derived from radiant heating is greater than by convection.
2. The radiant heat gives a greater feeling of warmth with a lower air temperature and this achieves about a 15 per cent saving in fuel costs.
3. In factories the lower air temperature gives a greater feeling of freshness, and production is known to increase.
4. The draughts are reduced to a minimum and dust is also kept down to a minimum.
5. Radiant heat does not heat the air through which it passes, but heats solid objects on which it falls, and so floors and walls derive warmth from the radiant heat rays. These warm surfaces set up convection currents, which reduce the heat lost from the human body by convection.

Chapter 4

Medium and high pressure hot water heating

Systems

In these systems the water is in a closed circuit and is subjected to pressure, by either steam or gas, so that its temperature may be raised above 100 °C. As the temperature of the water rises, the pressure in the system must also be raised to ensure that pressure is always above the evaporation pressure, otherwise the water would flash into steam and the systems would not function as high-temperature hot-water heating systems. The pressure on the water can be produced by steam maintained in the boiler, or a steam space inside an external drum. When a gas is used for pressurising, it is maintained in a pressure vessel and the boilers are completely filled with water. One of the simplest methods of pressurising a system is by means of a high-level head tank. This method however is limited, since few buildings are of suitable design. It would require a minimum height of 30 m above the highest main pipe in order to produce a pressure of 300 kPa.

Figure 4.1 shows a schematic diagram of a medium- or high-pressure system, using nitrogen inside a pressure cylinder. Nitrogen is preferable to air for this purpose because, unlike air, it is an inert gas, which prevents risk of corrosion. Furthermore it is less soluble in water than air.

Operation of the system

The pressure cylinder is maintained partly filled with water and partly with nitrogen. The boilers are fired and the expanded water enters the pressure cylinder until the nitrogen cushion is compressed to the working pressure of the system. At this pressure the pressure switch A is off, but if the pressure is exceeded the switch operates and opens the spill valve. Water is thus allowed to flow into the spill cistern and release the pressure in the cylinder to the correct level, when switch A closes the spill valve. If the pressure drops, due to leaks on packing glands, etc., the pressure switch B cuts in the feed pump, and water is forced into the cylinder until the correct water level and pressure is reached when switch B cuts out the pump.

Figure 4.2 shows a schematic diagram of a medium- or high-pressure boiler plant, utilising steam for pressurisation. The main flow and return pipes are taken below the water level in the boiler, so that steam cannot enter the pipework. In this system the steam and the water are at saturation temperature and any reduction of pressure on the water side will cause the water to flash into steam. To prevent this, a pipe is connected between the main flow and return pipes, so that cooled water from the return pipe can be allowed to mix with the hotter water in the flow pipe. A valve is fixed on this pipe to regulate the correct amount of mixing.

In another type of gas-pressurised system the cylinder is made large enough to accommodate all the expanded water without the need for a spill valve and cistern. The main reason behind the provision of a spill cistern however is the saving in cost of a large pressure cylinder. The circulation of the hot water is by means of a centrifugal pump, specially designed for use with high temperature water.

In both medium- and high-pressure systems it is essential that the feed water is treated, or otherwise scaling of boilers and equipment will occur. The workmanship and equipment must also be of a very high standard, with welded or flanged pipe joints, and bends must be of large radius and tees must be swept in the direction of flow. Allowances must be made for expansion and contraction of the pipework and air bottles or air valves fitted at the highest points of the circuits.

Table 4.1 shows the usual temperatures and pressures used for the systems.

Table 4.1   Temperatures and pressures used

Temperatures (°C)	Gauge pressure (kPa)	
150—200	600—1100	(high pressure)
120—135	300—400	(medium pressure)

Temperature control

The water temperature can be controlled at various points in the building by means of a three-way mixing valve. The valves allow cooler water in the return pipe to mix with the hotter water in the flow pipe at a controlled rate, before distribution to the various points in the building.

Types of heat emitters

If the water temperature exceeds 82 °C it is dangerous to install ordinary types of radiators that may be touched by occupants of the building. Above this temperature it is advisable to install overhead radiant strips or panels, unit heaters, convector type skirting heaters or fan convectors.

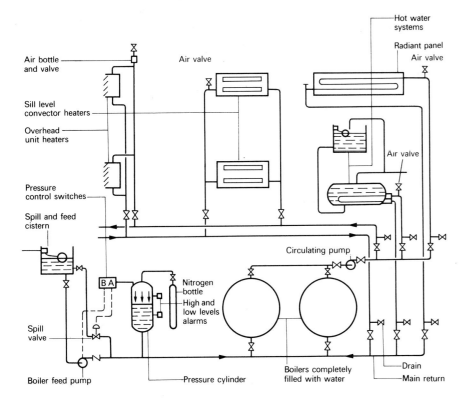

Fig. 4.1 Nitrogen pressurisation

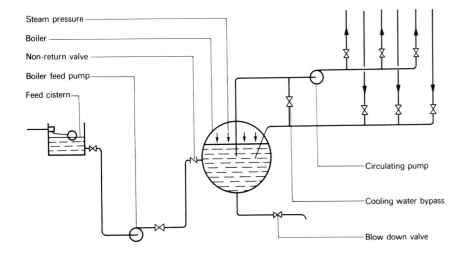

Fig. 4.2 Steam pressurisation

If it is required to have some part of a building heated by low-pressure hot water, which is easier to control by thermostats, non-storage type calorifiers may be used. These calorifiers have a battery of tubes inside which are heated by the medium- or high-pressure hot water. These tubes heat the water on the low-pressure side of the circuit, which is supplied with water from an open expansion and feed cistern.

Advantages over low-pressure systems

1. They contain a large thermal storage within the water, which will meet any sudden heat demands made upon the systems.
2. Because of the higher temperature of the water, smaller pipes and heat emitters may be used.
3. By use of smaller pipes, it is easier to distribute heat in very large buildings and less duct space is required.
4. The systems are very suitable for processing work in factories, district heating schemes and high-rise buildings.

Note: 1 Pascal = 1 N/m^2
         1 Bar   = 100 kPa

Steam heating

Steam as a heating medium is often used for the heating of industrial buildings, where the steam plant is also used for various processes requiring high temperatures. It is also used in hospitals, where it may provide a means of sterilising and of space heating and hot water supply indirectly by means of calorifiers. The system uses the latent heat of vaporation of water, which is equal to 2257 kJ/kg at atmospheric pressure, being greatly in excess of 419 kJ/kg, the sensible heat of water, to raise its temperature from 0° to 100 °C at the same pressure (see Table 4.2).

For this reason and because steam usually flows at a high velocity of between 24 to 36 m/s and at temperatures of between 100 °C and 198 °C, the pipework and heat emitters may be smaller than those used for hot water. Steam also flows under its own pressure and pumps are therefore not required. This is most useful for the heating of tall buildings, where the cost of pumping hot water is expensive. Steam systems, however, are more complicated than hot water and require more maintenance and supervision.

Terms relating to steam

Latent heat: heat which produces a change of state, without producing a change of temperature. In steam, it is the heat added to boiler water which converts some of the water into steam.

Sensible heat: heat which produces a rise in temperature of a substance without a change of state; in steam generation it is heat which has been added to water to raise its temperature to boiling point and over.

Saturated steam: steam which is generated in contact with the water.

Wet steam: steam which carries droplets of water in suspension.

Dry saturated steam: steam at the temperature of 100 °C which does not

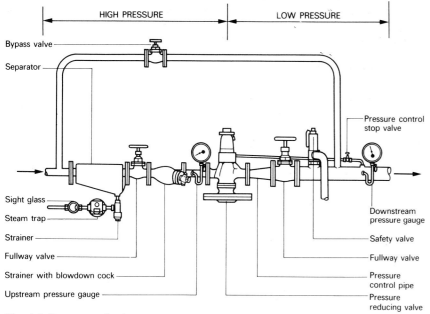

Fig. 4.3 Pressure-reducing set

Labels for Fig. 4.3:
- Bypass valve
- Separator
- HIGH PRESSURE
- LOW PRESSURE
- Pressure control stop valve
- Sight glass
- Steam trap
- Strainer
- Fullway valve
- Strainer with blowdown cock
- Upstream pressure gauge
- Downstream pressure gauge
- Safety valve
- Fullway valve
- Pressure control pipe
- Pressure reducing valve

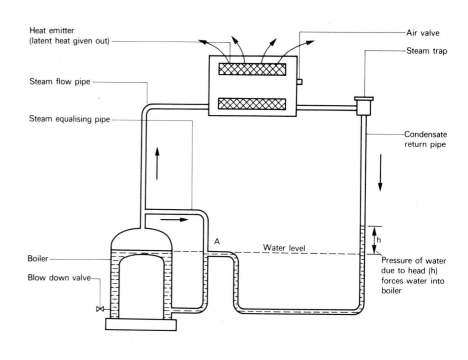

Fig. 4.4 Principle of steam heating

Labels for Fig. 4.4:
- Heat emitter (latent heat given out)
- Air valve
- Steam trap
- Steam flow pipe
- Steam equalising pipe
- Condensate return pipe
- Boiler
- Blow down valve
- A
- Water level
- h
- Pressure of water due to head (h) forces water into boiler

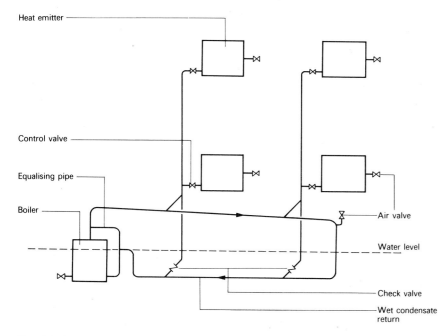

Fig. 4.5 One-pipe gravity system

Labels for Fig. 4.5:
- Heat emitter
- Control valve
- Equalising pipe
- Boiler
- Air valve
- Water level
- Check valve
- Wet condensate return

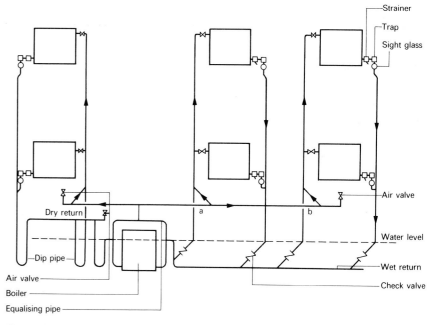

Fig. 4.6 Two-pipe gravity system

Labels for Fig. 4.6:
- Strainer
- Trap
- Sight glass
- Dry return
- a
- b
- Air valve
- Water level
- Wet return
- Check valve
- Dip pipe
- Air valve
- Boiler
- Equalising pipe

45

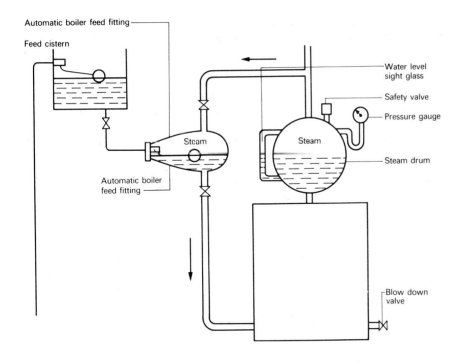

Fig. 4.7 Automatic boiler feed

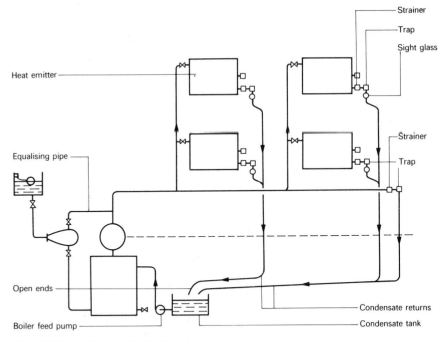

Fig. 4.8 Two-pipe mechanical system

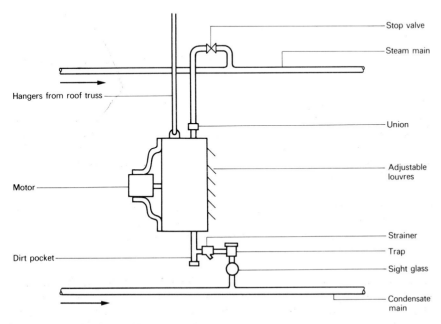

Fig. 4.9 Overhead unit heater connections

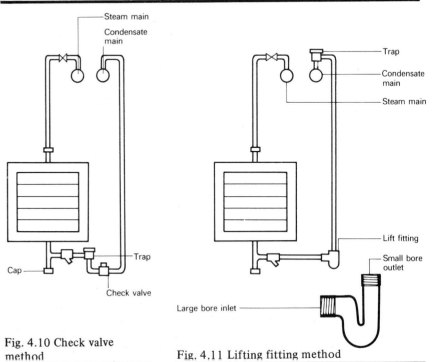

Fig. 4.10 Check valve method

Fig. 4.11 Lifting fitting method

contain any free droplets of water. It is the ideal type of steam, but is very rarely, if ever, found in practice.

Enthalpy, the total heat of steam: sensible heat plus latent heat.

Super-heated steam: steam which has further heat added, after it has left the boiler in which it was generated.

Absolute pressure: gauge pressure plus atmospheric pressure.

Steam pressures

Steam may be used at the following gauge pressures:

Low pressure up to 35 kPa

Medium pressure 140 to 550 kPa

High pressure 550 to 1400 kPa

Steam has a higher latent heat content at low pressure than at high pressure. Because latent heat is given out in the heat emitters and unless high temperatures are required, it follows that low-pressure steam is to be preferred. Low pressure also causes less risk of noise and wear on valves and other equipment.

In large installations having long steam mains, it may be necessary to use either medium or high pressure for the mains to overcome the frictional resistances, and in order to obtain low pressure a pressure-reducing set may be fitted at branch connections to various rooms or appliances. Figure 4.3 shows the equipment required to reduce the steam pressure and is known as a 'pressure-reducing set'.

Figure 4.4 shows the principle of operation of a simple steam heating system. The boiler is partly filled with water and, when cold, the remaining space in the boiler pipes and heat emitter is filled with air. When the water is heated to 100 °C steam is produced and flows up to the heat emitter, pushing the air before it. The air is allowed to escape through the air valve until the emitter is filled with steam, which on condensing gives up its latent heat. Water enters the steam trap which opens and allows the water to flow back to the boiler for reheating. Air is heavier than steam at low pressure and therefore it will be noticed that the air valve is fitted near the bottom of the heat emitter.

In a gravity system an equalising pipe is required, which causes the steam pressure to act downwards against the water pressure in the boiler to equalise the pressure at point A. The gradual increase in the height of water in the vertical condensate pipe, due to the formation of condensation, will overcome the steam pressure and allow water to enter the boiler.

Systems used

There are two systems:

1. *Gravity:* in which condensate runs back to the boiler by gravity.
2. *Mechanical:* in which condensate is pumped back to the boiler.

Figure 4.5 shows a one-pipe gravity system, where the flow pipe to the emitter is also used to carry condensate back to the boiler. The system does not require the use of steam traps, but may be noisy, due to water hammer.

Figure 4.6 shows a two-pipe gravity system, with steam traps connected to the outlets of each heat emitter. Water is prevented from being forced back up to the condensate return pipe by the steam pressure by fitting a check valve or a

dip on the pipe. Both methods are shown in the diagram. In order to check if the steam traps are functioning correctly, a sight flow indicator or a sight glass may be fitted to the outlet side of the traps, so that the operator can see if water is flowing and not steam.

The systems illustrated require the boiler to be initially filled by means of a force pump connected to the drain and blow down valve.

Figure 4.7 shows an automatic means of filling the boiler with water by means of a ball valve inside a sealed chamber, supplied from a cold-water feed cistern. A steel drum shown may be used to contain the steam space instead of the upper part of the boiler.

Figure 4.8 shows a two-pipe mechanical steam heating system, which requires a condensate tank or hot well. The tank may be sited at any level providing it is below the lowest heat emitter.

Types of heat emitters

In older systems radiators were used, but it has become more popular to use convector heaters, overhead unit heaters, radiant panels or strips.

Figure 4.9 shows the method of installing an overhead unit heater, with the steam main and condensate main at high and low levels respectively. If required, both mains may be at high level, providing either a check valve or a lifting fitting is installed, as shown in Figs 4.10 and 4.11.

Steam traps

The purpose of a steam trap is to remove the water which condenses inside appliances or pipelines. They may be divided into four groups, namely:

1. Thermostatic.
2. Ballfloat.
3. Bucket.
4. Thermodynamic.

Figures 4.12 4.13 4.14 4.15 and 4.16 show the various types of traps, which operate as follows:

Thermostatic type (see Fig. 4.12). The closed bellows contain a volatile spirit which has a boiling point to suit the temperatures involved. When steam enters the trap the volatile spirit expands and opens the bellows, thus closing the valve. When water enters the trap, which is at a lower temperature than steam, the spirit contracts and closes the bellows, thus opening the valve and allowing the water to flow back to the boiler.

Ballfloat type (see Fig. 4.13). When steam enters the trap the ballfloat is suspended and the weight of the float keeps the outlet valve closed. When water enters the trap the float becomes buoyant and opens the valve, thus allowing the water to flow back to the boiler.

Open bucket type (see Fig. 4.14). The outlet valve is closed as long as the bucket floats, but when water enters the trap it eventually overflows into the bucket, which causes the bucket to sink, thus opening the valve. Steam forces the water out of the bucket through the tube, until the bucket is buoyant, thus closing the valve.

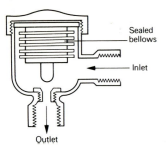

Fig. 4.12 Thermostatic
steam trap

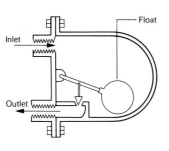

Fig 4.13 Ballfloat steam trap

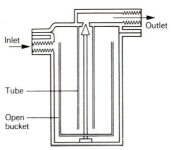

Fig. 4.14 Open bucket steam
trap

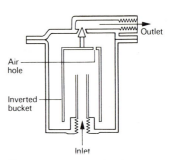

Fig 4.15 Inverted bucket
steam trap

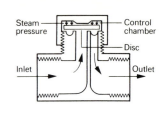

Fig 4.16 Thermodynamic
steam trap

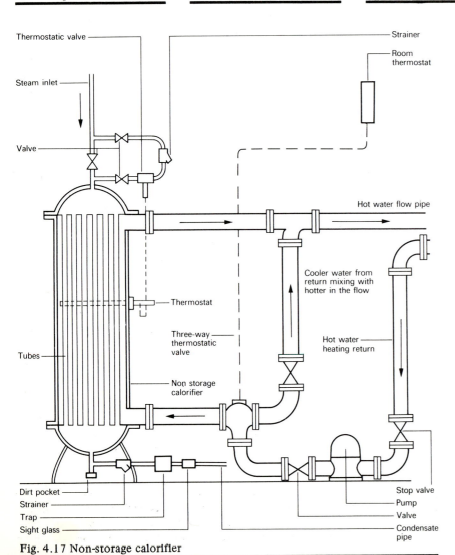

Fig. 4.17 Non-storage calorifier

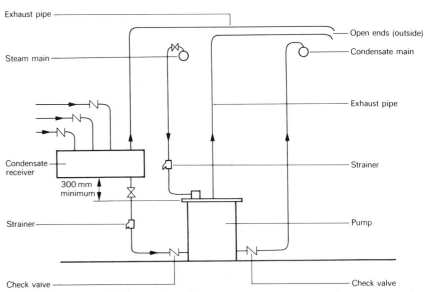

Fig. 4.18 Condensate lifting trap or pump

Fig. 4.19 Method of installing lifting trap or pump

Inverted bucket type (see Fig. 4.15). If steam enters the trap the bucket is lifted and the valve is closed. When water enters the trap the bucket eventually falls under its own weight and the valve opens, allowing the water to be forced out by the steam pressure.

Thermodynamic type (see Fig. 4.16). The trap operates on the Bernoulli principle which states that if no friction exists the total energy in a moving fluid is constant. The total energy is the sum of the kinetic pressure and potential energies of the moving fluid, so that an increase in one energy produces a decrease of another and vice versa.

When steam flows through the trap, an increase of kinetic energy is produced between the disc and the seating, which results in a reduction of pressure energy at this point and the disc moves nearer the seating until there is a reduction of kinetic energy. This reduction in kinetic energy produces an increase of pressure energy which tends to lift the disc from the seating, but is prevented from doing so by the steam pressure acting upon the top of the disc in the control chamber. Because the area at the top of the disc is greater than the area of the inlet underneath, the upper pressure forces the disc firmly on to its seat.

When water enters the trap the steam above the disc condenses, thus reducing the pressure, and the disc is forced up; this allows the water to flow through the trap. Water flows through the trap at a lower velocity than steam and does not cause sufficient reduction in pressure below the disc, so that the trap remains open until steam again enters. It will be clear that the trap operates on both kinetic energy and heat, hence the term, thermodynamic.

Non-storage calorifiers

These are used for providing hot water for space heating. The water in these calorifiers is heated indirectly by steam passing through a battery of pipes. To prevent overheating of the water, thermostatic control is required.

Figure 4.17 shows a method of installing a non-storage type calorifier, including details of thermostatic control of the calorifier and heating circuit.

Lifting trap

It may be necessary to lift condensate from low-level condensate branch pipes to a high-level condense main. This may be achieved by means of an automatic pump, as shown in Fig. 4.18 Figure 4.19 shows the method of installing the pump. The pump operates as follows:

1. Condensate enters the pump and forces air out of the chamber through the exhaust pipe.
2. The float rises to the top of the chamber and at this point closes exhaust valve A and opens the steam valve B.
3. Steam enters and forces the water out of the chamber to the main condensate pipe. On filling the chamber, check valve C is opened and D is closed. On emptying, check valve C is closed and D is opened.

Steam tables

These are used to find the properties at various pressures. Table 4.2 shows part of the steam tables, showing the values that may be obtained from them. A complete table may be obtained from the manufacturers of steam equipment. It will

Table 4.2 Steam tables

Gauge pressure (kPa)	Absolute pressure (kPa)	Temperature (°C)	Specific enthalpy			Specific volume
			Water	Evaporation	Steam	Steam (m³/kg)
			Sensible heat (kJ/kg)	Latent heat (kJ/kg)	Total heat (kJ/kg)	
—	5.0	32.88	137.82	2423.7	2561.5	28.192
—	10.0	45.81	191.83	2392.8	2584.7	14.674
—	15.0	53.97	225.94	2373.1	2599.1	10.022
—	20.0	60.06	251.40	2358.3	2609.7	7.649
—	25.0	64.97	271.93	2346.3	2618.2	6.204
—	30.0	69.10	289.23	2336.1	2625.3	5.229
—	35.0	72.70	304.30	2327.2	2631.5	4.530
—	40.0	75.87	317.58	2319.2	2636.8	3.993
—	45.0	78.70	329.67	2312.0	2641.7	3.580
—	50.0	81.33	340.49	2305.4	2645.9	3.240
—	55.0	83.72	350.54	2299.3	2649.8	2.964
—	60.0	85.94	359.86	2293.6	2653.5	2.732
—	65.0	88.01	368.54	2288.3	2656.9	2.535
—	70.0	89.95	376.70	2283.3	2660.0	2.365
—	75.0	91.78	384.39	2278.6	2663.0	2.217
—	80.0	93.50	391.66	2274.1	2665.8	2.087
—	85.0	95.14	398.57	2269.8	2668.4	1.972
—	90.0	96.71	405.15	2265.7	2670.9	1.869
—	95.0	98.20	411.43	2261.8	2673.2	1.777
—	100.0	99.63	417.46	2258.0	2675.5	1.694
0	101.3	100.00	419.04	2257.0	2676.0	1.673
5.0	106.3	101.40	424.9	2253.3	2678.2	1.601
10.0	111.3	102.66	430.2	2250.2	2680.4	1.533
15.0	116.3	103.87	435.6	2246.7	2682.3	1.471
20.0	121.3	105.10	440.8	2243.4	2684.2	1.414
25.0	126.3	106.26	445.7	2240.3	2686.0	1.361
30.0	131.3	107.39	450.4	2237.2	2687.6	1.312
35.0	136.3	108.50	455.2	2234.1	2689.3	1.268
40.0	141.3	109.55	459.7	2231.3	2691.0	1.225
45.0	146.3	110.58	464.1	2228.4	2692.5	1.186
50.0	151.3	111.61	468.3	2225.6	2693.9	1.149
55.0	156.3	112.60	472.4	2223.1	2695.5	1.115
60.0	161.3	113.56	476.4	2220.4	2696.8	1.083
65.0	166.3	114.51	480.2	2217.9	2698.1	1.051
70.0	171.3	115.40	484.1	2215.4	2699.5	1.024
75.0	176.3	116.28	487.9	2213.0	2700.9	0.997
80.0	181.3	117.14	491.6	2210.5	2702.1	0.971
85.0	186.3	117.96	495.1	2208.3	2703.4	0.946
90.0	191.3	118.80	498.9	2205.6	2704.5	0.923
95.0	196.3	119.63	502.2	2203.5	2705.7	0.901
100.0	201.3	120.42	505.6	2201.1	2706.7	0.881
105.0	206.3	121.21	508.9	2199.1	2708.0	0.860
110.0	211.3	121.96	512.2	2197.0	2709.2	0.841
115.0	216.3	122.73	515.4	2195.0	2710.4	0.823
120.0	221.3	123.46	518.7	2192.8	2711.5	0.806
125.0	226.3	124.18	521.6	2190.7	2712.3	0.788
130.0	231.3	124.90	524.6	2188.7	2713.3	0.773
135.0	236.3	125.59	527.6	2186.7	2714.3	0.757
140.0	241.3	126.28	530.5	2184.8	2715.3	0.743
145.0	246.3	126.96	533.3	2182.9	2716.2	0.728
150.0	251.3	127.62	536.1	2181.0	2717.1	0.714
155.0	256.3	128.26	538.9	2179.1	2718.0	0.701

Table 4.2 – *continued*

Gauge pressure (kPa)	Absolute pressure (kPa)	Temperature (°C)	Water Sensible heat (kJ/kg)	Evaporation Latent heat (kJ/kg)	Steam Total heat (kJ/kg)	Specific volume Steam (m³/kg)
160.0	261.3	128.89	541.6	2177.3	2718.9	0.689
165.0	266.3	129.51	544.4	2175.5	2719.9	0.677
170.0	271.3	130.13	547.1	2173.7	2720.8	0.665
175.0	276.3	130.75	549.7	2171.9	2721.6	0.654
180.0	281.3	131.37	552.3	2170.1	2722.4	0.643
185.0	286.3	131.96	554.8	2168.3	2723.1	0.632
190.0	291.3	132.54	557.3	2166.7	2724.0	0.622
195.0	296.3	133.13	559.8	2165.0	2724.8	0.612
200.0	301.3	133.69	562.2	2163.3	2725.5	0.603
205.0	306.3	134.25	564.6	2161.7	2726.3	0.594
210.0	311.3	134.82	567.0	2160.1	2727.1	0.585
215.0	316.3	135.36	569.4	2158.5	2727.9	0.576
220.0	321.3	135.88	571.7	2156.9	2728.6	0.568
225.0	326.3	136.43	574.0	2155.3	2729.3	0.560
230.0	331.3	136.98	576.3	2153.7	2730.0	0.552
235.0	336.3	137.50	578.5	2152.2	2730.7	0.544
240.0	341.3	138.01	580.7	2150.7	2731.4	0.536
245.0	346.3	138.53	582.8	2149.2	2732.0	0.529
250.0	351.3	139.02	585.0	2147.6	2732.6	0.522
255.0	356.3	139.52	586.9	2146.3	2733.2	0.515

Produced by permission of Spirax–Sarco Ltd, Charlton House, Cheltenham.

be clear from studying the table that the pressure of steam increases as the temperature increases and also specific latent heat increases as the pressure decreases. To find mass flow rate of steam passing through a pipe, the heat load is divided by the specific latent heat of steam.

Example 4.1. *Calculate the mass flow rate through a steam main, when the heat load is 2000 kW and the steam pressure is 100 kPa gauge.*

From the tables:

The specific latent heat of steam at 100 kPa gauge is 2201 kJ/kg

and the mass flow rate will be:

2000 ÷ 2201 = 0.9086 kg/s

In terms of volume, this mass flow rate will be:

0.9086 × 0.881 m³/kg = 0.8 m³/s

Feed water

The water used will require treatment to prevent corrosion and scaling of the boiler, pipework and equipment. A base exchange softening plant may be required, but the water will require partial re-hardening to prevent acidity which would otherwise cause corrosion.

Pipework

All steam and condensate pipes should be suitably insulated to prevent loss of heat. The pipework will require expansion joints, or loops, to relieve stresses due to expansion and contraction. All pipework must be provided with a fall of about 1 in 300, and at low points where condensate may be retained a drainage point must be fitted with a suitable steam trap.

District heating

District heating is an extension of the provision of space heating and hot water supply from a central boiler plant for one building, to the provision of these services from a central boiler plant for large development schemes, which may incorporate dwellings, offices, schools, factories and public buildings.

The term 'district' is a general term and covers the following types of heat distribution systems:

Block: a heat distribution system to a 'block' of similar buildings, such as a housing estate, an industrial estate, a shopping centre and offices.

Group: a heat distribution system to a group of the type of buildings described for the block system.

District: a heat distribution system for an entire city or town.

Figure 4.20 shows the distribution system for a group heating scheme.

The central station should, as far as possible, be sited near to the centre of the buildings being supplied with heat. This will reduce the frictional losses on the mains and also provide a better balanced distribution system. Although district heating can improve the environmental conditions in which people live and work, and also save fuel, the idea has not received the same consideration in this country as it has in many other countries. This is probably due to our temperate climate, plentiful supplies of cheap fuel and the high capital cost of a district heating installation.

With the development of new towns in Britain, there is now much greater interest being shown in district heating and there are now many schemes in this country. Nottingham has a scheme which uses combustible refuse to reduce the use of coal. Pimlico and Bankside of London have thermo-electric schemes which use waste heat from the steam turbines at Battersea and Bankside electric power stations. There are other schemes in London, and also at Billingham, Manchester, Sunderland, Oldham, Leicester and Paisley, to name just a few.

Bretton, a new town development by Peterborough Corporation, has a district heating scheme serving about 5000 homes, schools, factories, offices, library, cafés and health and social centres. The scheme uses natural gas and the system operates with low temperature hot water, at 83 °C during summer and 95 °C during winter. The system is pressurised by applying nitrogen gas to the water in a pressure vessel.

The USA pioneered district heating and there are schemes in New York, Washington, Detroit, Philadelphia, Cleveland, Chicago and many other cities. In Europe, Manchester installed the first district heating scheme and there are now schemes in Germany, Norway, Denmark, Sweden, Holland, Belgium, France and Italy.

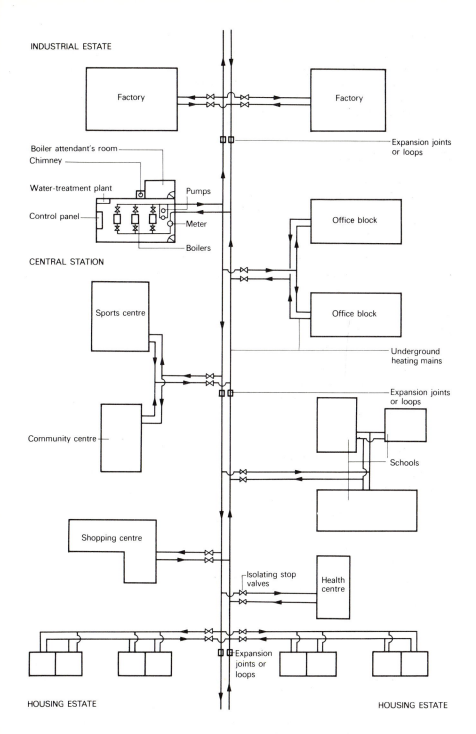

INDUSTRIAL ESTATE

Factory

Factory

Expansion joints or loops

Boiler attendant's room

Chimney

Water-treatment plant

Pumps

Office block

Control panel

Meter

Boilers

CENTRAL STATION

Sports centre

Office block

Underground heating mains

Community centre

Expansion joints or loops

Schools

Shopping centre

Isolating stop valves

Health centre

Expansion joints or loops

HOUSING ESTATE

HOUSING ESTATE

Fig 4.20 Group heating scheme

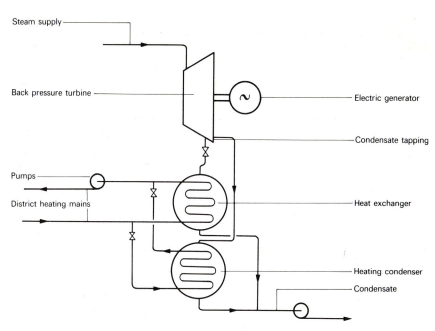

Steam supply

Back pressure turbine

Electric generator

Condensate tapping

Pumps

District heating mains

Heat exchanger

Heating condenser

Condensate

Fig 4.21 Back pressure steam turbine and district heating

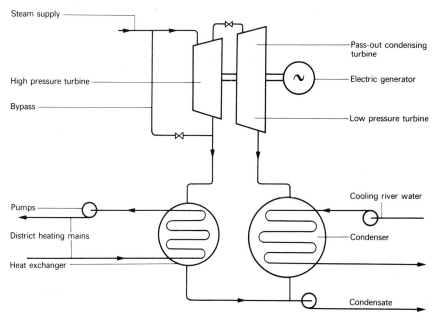

Steam supply

Pass-out condensing turbine

High pressure turbine

Electric generator

Bypass

Low pressure turbine

Pumps

Cooling river water

District heating mains

Condenser

Heat exchanger

Condensate

Fig 4.22 Pass-out condensing turbine and district heating

51

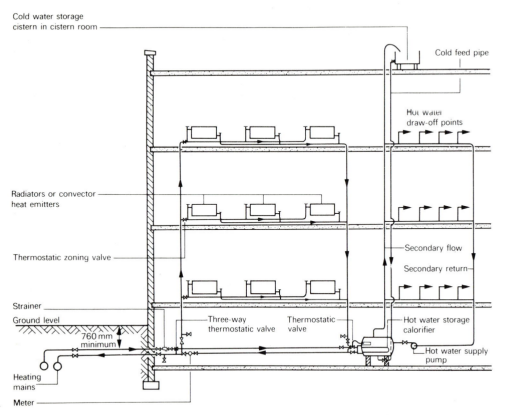

Fig 4.23 Two-pipe system to heating and hot water supply for a large building

Labels (Fig 4.23):
- Cold water storage cistern in cistern room
- Cold feed pipe
- Hot water draw-off points
- Radiators or convector heat emitters
- Thermostatic zoning valve
- Secondary flow
- Secondary return
- Strainer
- Ground level
- 760 mm minimum
- Three-way thermostatic valve
- Thermostatic valve
- Hot water storage calorifier
- Hot water supply pump
- Heating mains
- Meter

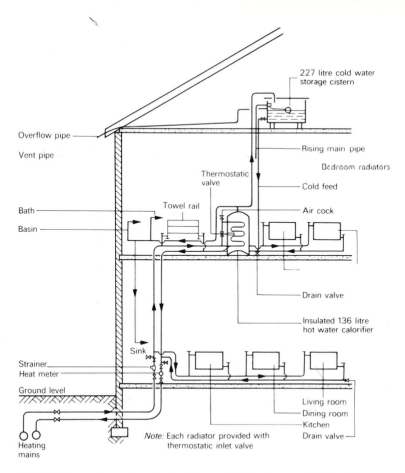

Fig 4.24 Two-pipe system to heating and hot water supply for a house

Labels (Fig 4.24):
- 227 litre cold water storage cistern
- Overflow pipe
- Vent pipe
- Rising main pipe
- Bedroom radiators
- Thermostatic valve
- Cold feed
- Bath
- Towel rail
- Air cock
- Basin
- Drain valve
- Insulated 136 litre hot water calorifier
- Sink
- Strainer
- Heat meter
- Ground level
- Living room
- Dining room
- Kitchen
- Drain valve
- Heating mains
- Note: Each radiator provided with thermostatic inlet valve

Advantages of district heating can be summarised as follows:

1. An electric power station operates at only about 35 per cent efficiency and about 60 per cent of the heat input is wasted through the turbine exhaust. This waste heat may be used for supplying heat for buildings, within a reasonable distance of the power station.

Figure 4.21 shows a simplified arrangement of steam back-pressure turbine and district heating circuit and Fig. 4.22 shows a simplified arrangement of a pass-out condensing turbine and district heating. In the latter type, a greater quantity of electrical energy is produced, but the total vacuum steam, passed to waste, can no longer be used for heating purposes and is therefore condensed and removed by river water.

2. Domestic refuse can be used, and since the weight of refuse produced is increasing at the rate of about 3.5 per cent per annum this would be an excellent answer to the disposal of this otherwise waste material.

3. With only one boiler room, fuel store and chimney, there is a large saving in space and also construction costs.

4. A few boilers, pumps and other equipment can be maintained at peak efficiency.

5. There is very little air pollution from one boiler room and chimney.

6. Fewer boiler attendants are required and therefore there is a saving in manpower.

7. There is a reduction in the cost to the consumer of providing heating and hot water, and there is less risk of fire in buildings.

8. It is possible to obtain cheaper fuel supplies due to bulk purchasing and cheaper transport to only one boiler room.

9. There is a greater incentive for local authorities to provide central heating for old people's dwellings and this would reduce the risk of deaths in winter due to hyperthermia.

10. There is a reduction in the amount of service roads for the transport of fuel.

11. There is less worry to the consumer in maintaining the heating and hot water equipment.

12. Because the water used for the system is treated, there is a large reduction in the problem of scaling and rusting of boilers, heat emitters and pipework.

13. There are no house or factory chimneys to spoil the landscape. The only chimney is at the central station and this may be hidden from view by the discreet siting, or by installing the chimney, in a nearby office block.

Heat distribution

In this country the medium used for the supply of heat is hot water pumped through a distribution main (some countries, including the USA, use steam as the heat medium). The water is usually at high or medium pressure of 300 to 1000 kPa and a temperature of 120° to 150 °C. Low-pressure systems operating at about 150 kPa and 80° to 100 °C have also been used.

The hot water is circulated through a number of insulated mains, each of which serves an individual consumer, or a sector of the district. Industrial and other large consumers take their heating requirements direct from the mains and storage type calorifiers are installed, which are thermostatically controlled, to provide hot water at 60 °C. Domestic and smaller consumers may also use this arrangement, or alternatively, both the heating and hot water supply may be provided indirectly from thermostatically controlled calorifiers.

There are three types of heat distribution systems, namely:

Two-pipe: this is the most common type of system and the heating mains serve both space heating and hot water supply. The power used for the pumps may be reduced during the summer months by using variable speed drives, or by using smaller pumps than are used in the winter months.

Figure 4.23 shows a two-pipe heat distribution system for a large building, such as a hospital, office block, factory or a university.

Figure 4.24 shows a two-pipe system for a small house.

Three-pipe system: this has both a small diameter and a large diameter flow main and a common large diameter return main. During summer, when only hot water is required, the small diameter flow pipe and the large diameter return pipe are used, which will reduce the pumping and heat losses on the large diameter flow main, which is only used during winter.

Four-pipe system: this system has both heating and hot water flow and return mains, which save the cost of individual hot water calorifiers for each building, but increase the cost of the distribution mains. This system is often used for a block scheme.

Figures 4.25 and 4.26 show schematic diagrams of the three-pipe and four-pipe distribution systems respectively, and Fig. 4.27 shows a plan of a central plant station for a block scheme, using the four-pipe distribution system.

Figure 4.28 shows a section through an oil tank storage room and the method of installing the oil tank. The boilers may be fired by solid fuel or natural gas.

Figure 4.29 shows the method used to supply the boilers with solid fuel and Fig. 4.30 shows a section through an 'economic' boiler, which is often used for district heating.

Charges for heat supply

The consumers may be charged by the amount of heat passing through a heat meter installed inside the building, or by a flat rate charge, based usually upon the number of occupants, or the number and size of rooms in the building. It may be necessary to meter the heat supply to large consumers, such as offices, factories and schools, and charge a flat rate for domestic consumers.

The charge for heat made by the amount passing through a meter has the advantage of providing a means of using the heat more economically. Some domestic consumers may require less heat, due to being out at work during the day, or requiring low air temperatures. Both domestic and large consumers are encouraged to economise in the use of heat when a meter is installed. Meters, however, add to the capital and maintenance cost of the installation and also require reading, which increases the administrative costs. It is also sometimes difficult to gain entry to dwellings when people may be out at work, but this problem may be solved by installing the meter in a locked cupboard, in which the door can be opened by the meter reader and the readings taken from outside the building. Meters should always be installed in the central station, to record the heating output from the boilers.

Distribution mains

The success of a district heating system depends largely upon the correct installation of the heating mains. It is essential that adequate provision is made for expansion and contraction of the pipework, and the pipe joints should be thoroughly made and tested. The type of ductwork for the pipes and method of thermal insulation are important considerations. It is also essential that the pipe insulation is kept dry, or otherwise its efficiency would be seriously impaired and therefore, if possible, the pipe duct should be laid above the ground water table and drainage below the duct pipes should be laid below the duct. These drainage pipes should be carried to a soakaway or a pumping sump.

Various methods are used for insulating the pipes against heat losses and the type used will depend upon the size of the heating mains, cost and speed of laying, accessibility and type of ground.

Aerated concrete-filled duct

A pre-cast concrete duct is placed in the excavated trench and the heating pipes are then placed on to concrete blocks, which rest on the base of the concrete duct. After welding and testing, the pipes are wrapped with corrugated cardboard, which prevents bonding between the pipes and the insulation. The duct is then filled with aerated concrete slurry, which surrounds the heating pipes. The insulation is completed by placing a pre-cast concrete cover over the duct and backfilling the trench.

Figure 4.31 shows a cross section through the concrete duct and the method of insulating the pipes.

Arched duct

When the heating mains are to be laid below roads where there is a possibility of heavy traffic, an arched duct will provide greater strength than the rectangular duct. The duct is easily laid and instead of filling with cellular concrete the pipes are insulated with glass fibre or asbestos.

Figure 4.32 shows the arched conduit method, with the heating pipes supported on rollers and guides to allow for expansion and contraction.

Conduit casing

Since ground water is one of the main problems of laying the underground

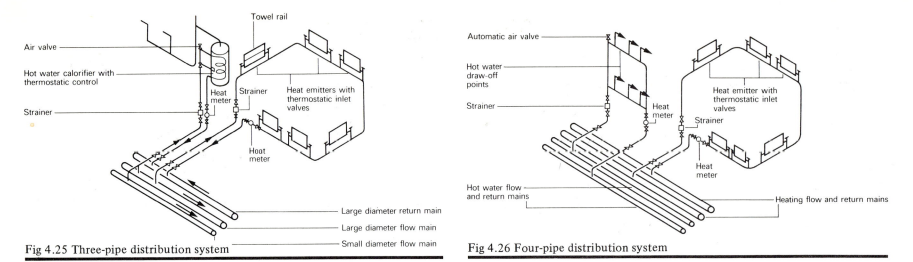

Fig 4.25 Three-pipe distribution system

Air valve

Hot water calorifier with thermostatic control

Strainer

Towel rail

Heat meter

Strainer

Heat emitters with thermostatic inlet valves

Heat meter

Large diameter return main

Large diameter flow main

Small diameter flow main

Fig 4.26 Four-pipe distribution system

Automatic air valve

Hot water draw-off points

Strainer

Heat meter

Heat emitter with thermostatic inlet valves

Strainer

Heat meter

Hot water flow and return mains

Heating flow and return mains

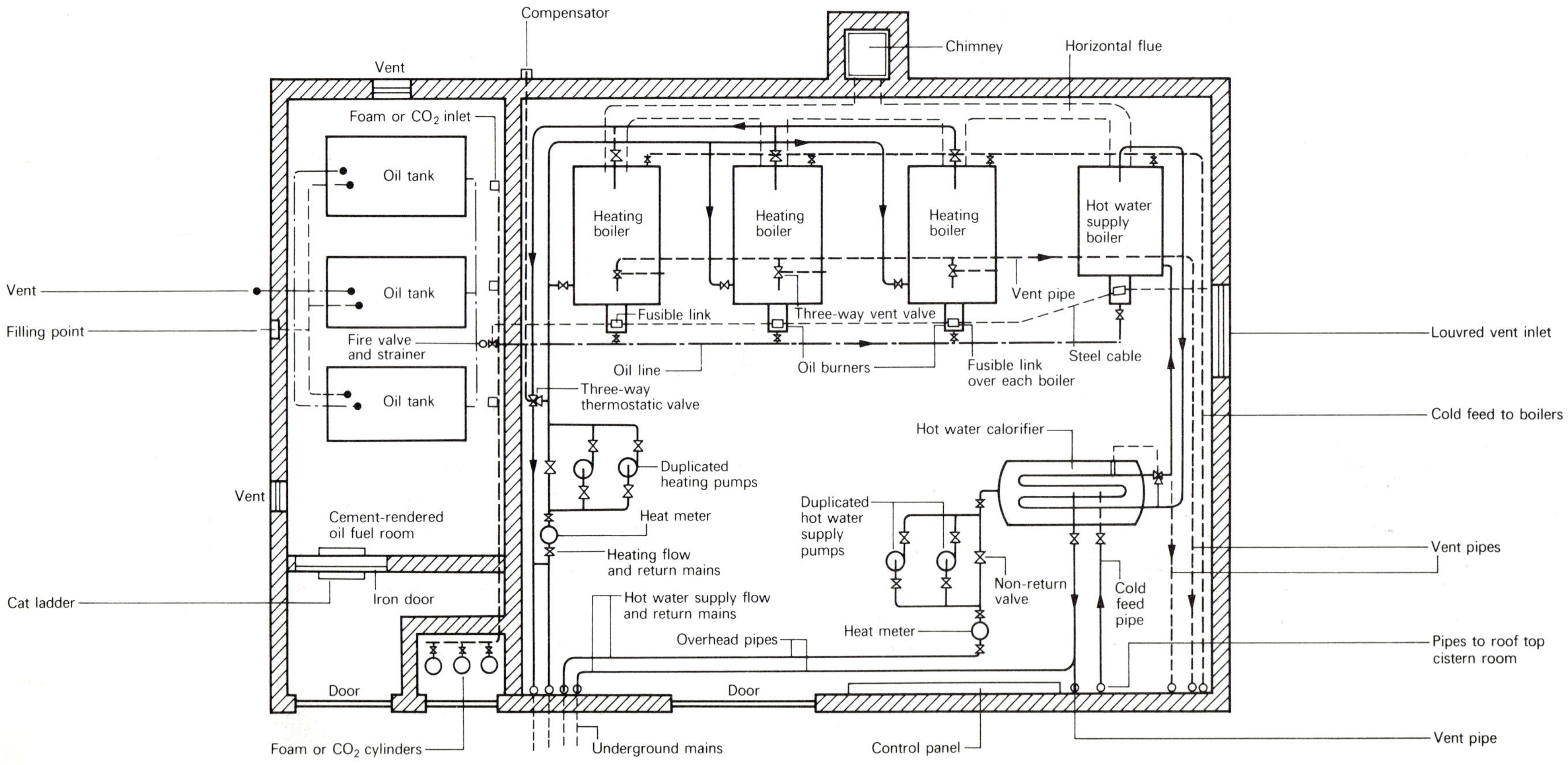

Fig 4.27 Central plant station for a small block scheme

Compensator

Chimney

Horizontal flue

Vent

Foam or CO$_2$ inlet

Vent

Filling point

Vent

Cement-rendered oil fuel room

Cat ladder

Door

Foam or CO$_2$ cylinders

Oil tank

Oil tank

Oil tank

Fire valve and strainer

Iron door

Three-way thermostatic valve

Heating boiler

Heating boiler

Heating boiler

Hot water supply boiler

Fusible link

Three-way vent valve

Oil line

Oil burners

Fusible link over each boiler

Steel cable

Vent pipe

Louvred vent inlet

Cold feed to boilers

Hot water calorifier

Duplicated heating pumps

Heat meter

Heating flow and return mains

Duplicated hot water supply pumps

Non-return valve

Vent pipes

Hot water supply flow and return mains

Overhead pipes

Heat meter

Cold feed pipe

Pipes to roof top cistern room

Door

Underground mains

Control panel

Vent pipe

54

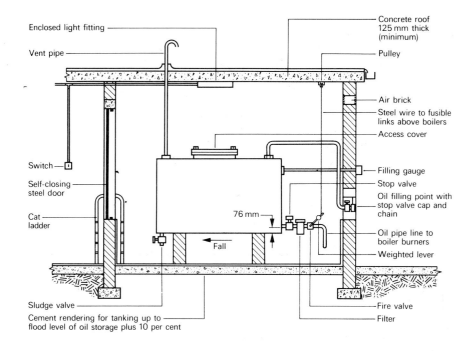

Fig 4.28 Detail of oil tank room

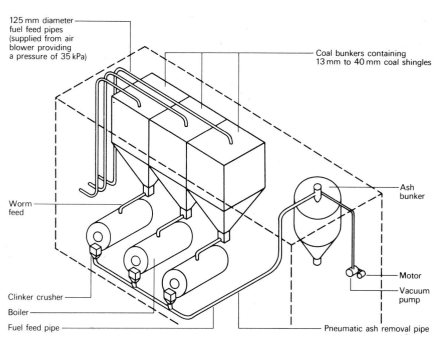

Fig 4.29 Solid fuel boilers

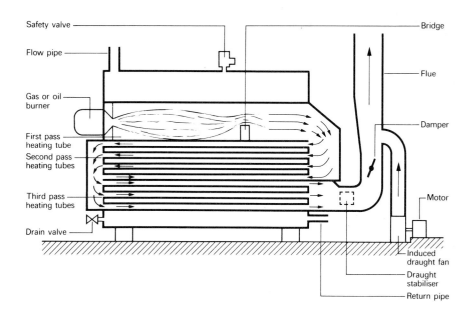

Fig 4.30 Economic boiler

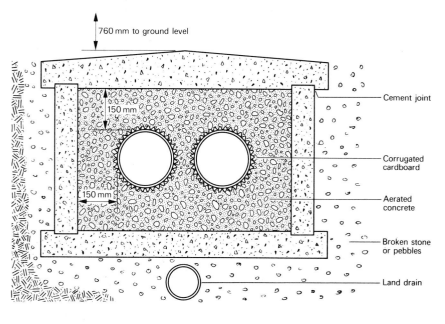

Fig 4.31 Aerated concrete fill

55

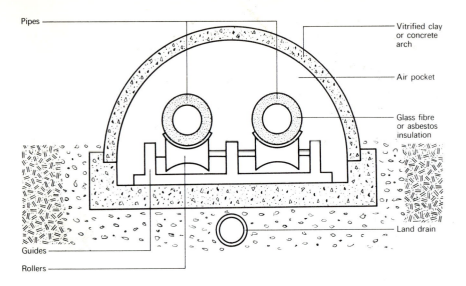

Pipes

Vitrified clay
or concrete
arch

Air pocket

Glass fibre
or asbestos
insulation

Land drain

Guides

Rollers

Fig 4.32 Arched conduit

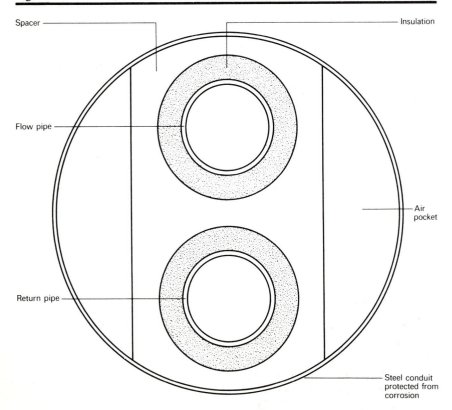

Spacer

Insulation

Flow pipe

Air
pocket

Return pipe

Steel conduit
protected from
corrosion

Fig. 4.33 Steel pipe conduit

distribution mains, the heating pipes can be fitted inside a watertight pipe conduit casing.

Figure 4.33 shows a section through a pipe conduit and the heating mains. The conduit and heating pipes are manufactured as complete units in 12 m lengths. The normal method of installation is to weld three lengths together above the trench and hydraulically testing their line. This section of the pipework is then lowered into the trench and this practice is continued along the full length of the line. The conduit joints are given a protective coating before the trench is backfilled. This method of pipe laying ensures a quick installation, with trenches open for minimum periods, thus reducing labour costs and the inconvenience of open trenches.

The pipe conduit casing is protected from corrosion by layers of coal tar, reinforced with glass fibre and an outer layer of coal tar-saturated asbestos felt. The conduit should be provided with accessible drain plugs for emergency use in the event of a leak on the pipework.

Chapter 5

Pipe-sizing for heating pump duty and micro-bore systems

Pumped circulation

Pumped circulation is used for all medium and large installations and has the following advantages over natural or gravity circulation:

1. Quicker response to heating of the building.
2. Circulation is independent of the temperature difference between the flow and return water.
3. Smaller pipes may be used, thus saving in capital costs, saving in space, and neater in appearance.
4. The boiler room may be sited on the roof.
5. Any type of heat emitter may be used.

Although the cost of the pump or pumps and the cost of power for running them must be considered, the use of pumped circulation adequately compensates by the saving in the overall capital and running cost of the installation.

The following steps may be taken in finding the sizes of the heating pipes:

1. Calculate the heat requirements for each room as described in Chapter 1.
2. Make a sketch of the system showing the positions of the heat emitters, boiler and the pipe runs. At this stage it will be necessary to decide on whether the system is to be one- or two-pipe and also the number of circuits to serve the various rooms in which heat emitters are to be installed.
3. Decide on the temperature of the flow and return water, i.e. the temperature drop. Although this may appear arbitrary, the diameter of the pipe and pump duty found later will provide for this temperature drop.
4. Estimate the heat emission to be expected from the pipes. It is usual to allow between 5 and 30 per cent of the total heat from the emitters for the heat emission from the pipes. The amount will depend upon the diameter and length of pipes.
5. Estimate the mass rates of flow through the circuit or circuits in kg/s.
6. Decide upon the velocity of water flow.
7. From the pipe-sizing table find a suitable size pipe or pipes.
8. Measure the actual lengths of the circuit or circuits and to this add an allowance for frictional losses due to fittings, to obtain the effective pipe length.
9. Find the pressure required for the pump.

An example will show how these steps are carried out.

Example 5.1 *Figure 5.1 shows a single-pipe circuit. Assuming each radiator has a heat emission of 8 kW and the temperature drop across the system is 15 °C, find by the use of Table 5.1 the approximate diameter of a steel pipe for the main and the pump pressure. Use a velocity of flow of water of 1 m/s.*

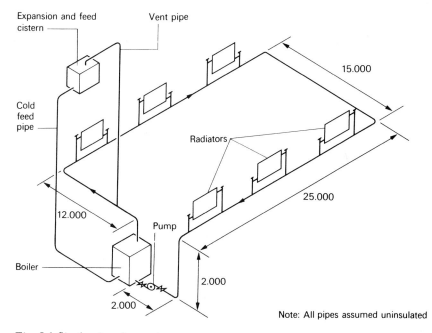

Fig. 5.1 Single-pipe ring system

The single-pipe circuit must carry sufficient energy in the water flowing through it to supply the total heat required for the radiators, plus the heat required for the pipes.

total heat emission from radiators =	48 kW
heat emission from the pipes assuming 20 per cent of the heat emission from radiators	=	9.6 kW
total	=	57.6

Mass flow rate (kg/s)

$$\text{power} = \text{kg/s} \times \text{s.h.c.} \times (t_f - t_r) \quad \text{(kW)}$$

$$\therefore \text{flow rate} = \frac{\text{kW}}{\text{s.h.c.} \times (t_f - t_r)} \quad \text{(kg/s)}$$

where kW = total heat carried by the pipe
s.h.c. = specific heat capacity of water taken as 4.2 kJ/kg°C
t_f = temperature of flow water °C
t_r = temperature of return water °C

$$\text{flow rate} = \frac{57.6}{4.2 \times 15}$$

$$= 0.92 \text{ kg/s}$$

Table 5.1 Flow of water at 75°C in black steel pipes. Medium-grade pipe to BS 1387:1967

20 mm			25 mm		32 mm		40 mm		50 mm			Pressure drop (Pa per m run of pipe)
V	M	EL	M	EL	M	EL	M	EL	M	EL	V	
	0.122	0.7	0.228	1.0	0.480	1.4	0.720	1.8	1.35	2.4		80.0
	0.124	0.7	0.232	1.0	0.488	1.4	0.732	1.8	1.37	2.4		82.5
	0.126	0.7	0.236	1.0	0.496	1.4	0.743	1.8	1.39	2.4		85.0
	0.128	0.7	0.240	1.0	0.503	1.4	0.755	1.8	1.41	2.4		87.5
	0.130	0.7	0.243	1.0	0.511	1.4	0.766	1.8	1.43	2.4		90.0
	0.132	0.7	0.247	1.0	0.518	1.5	0.778	1.8	1.45	2.4		92.5
	0.134	0.7	0.251	1.0	0.526	1.5	0.789	1.8	1.48	2.4		95.0
	0.136	0.7	0.254	1.0	0.533	1.5	0.800	1.8	1.50	2.4		97.5
	0.138	0.7	0.258	1.0	0.540	1.5	0.810	1.8	1.52	2.4		100.0
	0.152	0.7	0.284	1.0	0.595	1.5	0.893	1.8	1.67	2.4		120.0
	0.165	0.8	0.308	1.0	0.646	1.5	0.968	1.8	1.81	2.5		140.0
0.5	0.178	0.8	0.331	1.0	0.693	1.5	1.04	1.8	1.94	2.5		160.0
	0.189	0.8	0.353	1.0	0.738	1.5	1.11	1.8	2.06	2.5	1.0	180.0
	0.200	0.8	0.373	1.0	0.780	1.5	1.17	1.9	2.18	2.5		200.0
	0.211	0.8	0.392	1.1	0.820	1.5	1.28	1.9	2.29	2.5		220.0
	0.221	0.8	0.411	1.1	0.858	1.5	1.29	1.9	2.40	2.5		240.0
	0.230	0.8	0.428	1.1	0.895	1.5	1.34	1.9	2.50	2.5		260.0
	0.239	0.8	0.445	1.1	0.931	1.5	1.39	1.9	2.60	2.6		280.0
	0.248	0.8	0.462	1.1	0.965	1.5	1.44	1.9	2.69	2.6		300.0
1.0	0.257	0.8	0.478	1.1	0.998	1.6	1.49	1.9	2.78	2.6	1.5	320.0

V = velocity of flow m/s
M = mass rate of flow kg/s
EL = equivalent length factor

From the pipe-sizing table (Table 5.1), a 40 mm diameter steel pipe will carry 0.968 kg/s at a velocity of flow of 1 m/s giving a pressure loss per metre run of 140 Pa and an EL factor of 1.8.

Check on the estimated heat loss on the pipes of 20 per cent

Table 5.2 shows that for a 50°C difference between the pipe surface and the surrounding air, the heat loss on a 40 mm diameter pipe is 105 watts per metre run.

58

Table 5.2 Theoretical heat emission from single horizontal steel pipes freely exposed in ambient air

Nominal bore (mm)	Temperature difference of surface to surroundings (°C)				
	40	45	50	55	60
8	28	32	35	39	43
10	33	37	41	46	50
15	40	46	53	59	66
20	48	56	62	70	78
25	58	68	78	88	98
32	71	82	93	110	120
40	78	92	105	110	130
50	96	112	130	150	170
65	120	140	160	180	200
80	140	160	180	211	230

Heat emission (W/m run)

Since the total length of pipe is 82 m,

heat emission from pipe = 82 × 105 = 8610 watts

$$\text{percentage emission} = \frac{8610}{48\,000} \times \frac{100}{1} = 18 \text{ per cent}$$

This is only a 2 per cent difference between the estimated heat loss from the pipes and the actual heat loss, and therefore the pipe diameter is suitable.

Pump duty

Before the pressure required by the pump can be determined, it will be necessary to find the effective length of the pipe. The effective length is the addition of the equivalent length due to the resistances of fittings and the actual length.

The resistances of the fittings are computed, in terms of equivalent length of straight pipe of corresponding diameter, by the product of an equivalent length factor found from Table 5.1 and a velocity pressure-loss factor, K.

Table 5.3 gives the K factor for some of the common types of fittings.

The pressure created by a pump on a one-pipe system does not force water through the radiators (unless special injector tees are used). The hot water therefore circulates through the radiators by natural convection created by the difference in density of the water flowing into the radiators and the cooler, denser water flowing out.

The frictional resistances through the tees for the purpose of determining the effective length of pipe will therefore be ignored.

Equivalent length of pipe = number of fittings × K × EL.

From Table 5.3, the K factor for a boiler is 5; and for 40 mm diameter

Table 5.3 Values of velocity pressure factor K for pipe fittings and equipment

Tees	K factor
	0.5 plus 0.2 for enlargement or reduction
Gate valve	0.2
Angle valve	0.5
Radiator	5.0
Sectional boiler	5.0
90° elbow 10 mm to 25 mm	0.8
90° elbow 32 mm to 50 mm	0.7
90° elbow 65 mm to 90 mm	0.6
90° bend 10 mm to 25 mm	0.7
90° bend 32 mm to 50 mm	0.5
90° bend 65 mm to 90 mm	0.4

Reductions or enlargements on a straight run of pipe

3 : 2 reduction	0.3
2 : 1 reduction	0.4
3 : 1 reduction	0.4
4 : 1 reduction	0.5
3 : 2 enlargement	0.4
2 : 1 enlargement	0.7
3 : 1 enlargement	0.9

90° bends, it is 0.5. The equivalent lengths are therefore:

$$\text{boiler} = 1 \times 5 \times 1.8 = 9.0 \text{ m}$$
$$\text{bends} = 6 \times 0.5 \times 1.8 = 5.4 \text{ m}$$
$$14.4 \text{ m}$$

$$\text{actual length} = 1 + 12 + 25 + 15 + 25 + 2 + 2 = 82 \text{ m}$$
$$\therefore \text{ effective length of pipe} = 82 + 14.4 = 96.4 \text{ m}$$

Pump pressure

The pressure loss through the pipe from Table 5.1 was found to be 140 Pa per metre run of pipe.

$$\therefore \text{ total pressure loss} = 96.4 \times 140 = 13\,496 \text{ Pa}$$

This is approximately equal to 13.5 kPa pressure and is equal to $13.5 \div 9.81 = 1.4$ m head of water.

Pumps are ordered giving the pressure developed and the flow rate in kg/s or litre/s. A pump providing a pressure of 13.5 kPa and giving a mass flow rate of 1 kg/s would be satisfactory.

Two-pipe system

In a two-pipe system, the flow pipes will reduce in diameter as they pass along the circuit, and the return pipes will increase in diameter as they return to the boiler.

The pressure created by a pump in a two-pipe system acts through the heat emitters and therefore convector heaters, which offer more resistance to the flow of water than radiators, can be used. The frictional resistances of the tees will, however, have to be taken into account when determining the effective length of pipe and the pump pressure.

Index circuit

In order to determine the pump pressure, the index circuit has to be selected. This is the circuit having the greatest resistance to the flow of water and supplies the index heat emitter. For a pumped system the index circuit is always taken to be the longest circuit.

Heat distribution

Unlike the one-pipe system, where the cooler water from the heat emitter passes back into the flow pipe, in the two-pipe system, the cooler water from each heat emitter passes into the return pipe and this provides a better balance of heat distribution in the system.

Example 5.2 *Figure 5.2 shows a two-pipe parallel system to be used as an alternative to the one-pipe ring system given in example 5.1. Using the follow-data and assumptions find by use of Tables 5.1–5.3 the approximate diameters of the flow and return mains and pump duty.*

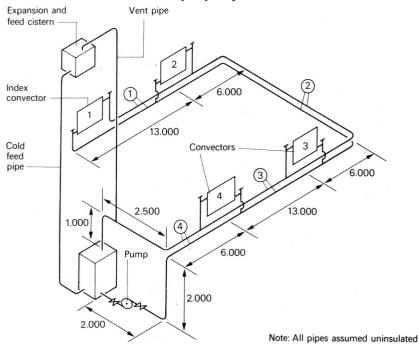

Fig. 5.2 Two-pipe parallel system

1. *Heat emission from each convector* $= 12\,kW$
2. *Temperature drop across the system* $= 15\,^{\circ}C$
3. *Mean temperature of water* $= 72\,^{\circ}C$
4. *Specific heat capacity of water* $= 4.2\,kJ/kg\,^{\circ}C$
5. *Velocity of flow of water* $= 1\,m/s$

Pipe sizing

Before the pipe-sizing procedure is commenced, a reference number or letter should be allocated to each section of pipework for identification purposes. The pipe-sizing may be started from the boiler or the farthest convector from it. In this example, the index convector is used as the starting point.

In order to find the mass flow rate through each section of pipework, an estimate is made of the carrying capacity of the pipes for each section, by taking the heat required for the convector and adding an estimated percentage for the pipework heat emission. This estimate can be checked and the pipe-sizing revised, if required.

Pipes number 1: Heat emission from convector 1 and pipework 1

emission from convector = 12.0 kW
emission from pipework = 2.4 kW
5 per cent of 48 kW

total = 14.4 kW

$$\text{flow rate} = \frac{14.4}{4.2 \times 15}$$

$$= 0.228 \text{ kg/s}$$

From pipe-sizing Table 5.1 a 20 mm diameter steel pipe will carry 0.257 kg/s giving a pressure loss per metre run of 320 Pa and an EL factor of 0.8.

Check on estimated heat loss of 5 per cent assuming a temperature of pipe surface to the ambient air of $50\,^{\circ}C$ (see Table 5.2).

26 m × 62 = 1612 watts

$$\text{percentage emission} = \frac{1612}{48\,000} \times \frac{100}{1}$$

heat loss = 3.4 per cent

Since 5 per cent was allowed, pipes number 1 are satisfactory.

Pipes number 2: Emission from convector 2 and pipework 2

emission from convector = 12.0 kW
emission from pipework = 4.8 kW
10 per cent of 48 kW

total = 16.8 kW

$$\text{flow rate} = \frac{16.8 \text{ kW}}{4.2 \times 15}$$

$$= 0.267 \text{ kg/s}$$

Pipes number 2 will also have to carry the heat required for emitter and pipework 1.

total flow rate = 0.228 + 0.267

$$= 0.5 \text{ kg/s}$$

From the pipe-sizing Table 5.1 , a 32 mm diameter pipe will carry 0.503 kg/s giving a pressure loss per metre run of 87.5 Pa and an EL factor of 1.4.

Check on estimated heat loss of 10 per cent (see Table 5.2).

54 m × 93 = 5022 watts

$$\text{percentage heat loss} = \frac{5022}{48\,000} \times \frac{100}{1}$$

heat loss = 10.46 per cent

Pipes number 2 are therefore satisfactory for practical purposes.

Pipes number 3: Emission from convector and pipework 3

emission from convector = 12.0 kW
emission from pipework = 2.4 kW
5 per cent of 48 kW

total = 14.4 kW

$$\text{flow rate} = \frac{14.4}{4.2 \times 15}$$

$$= 0.228 \text{ kg/s}$$

Pipes number 3 will also have to carry the heat required for emitters and pipework 1 and 2.

total flow rate = 0.5 + 0.228

$$= 0.728 \text{ kg/s}$$

From pipe-sizing Table 5.1 a 32 mm diameter pipe will carry 0.738 kg/s giving a pressure loss per metre run of 180 Pa and an EL factor of 1.5.

Check on estimated heat loss of 5 per cent (see Table 5.2).

26 m × 93 = 2418 watts

$$\text{percentage heat loss} = \frac{2418}{48\,000} \times \frac{100}{1}$$

heat loss = 5 per cent

Pipes number 3 are therefore satisfactory.

Pipes number 4: Emission from convector 4 and pipework 4

emission from convector = 12.0 kW
emission from pipework = 2.4 kW
5 per cent of 48 kW

total = 14.4 kW

$$\text{flow rate} = \frac{14.4}{4.2 \times 15}$$

$$= 0.228 \text{ kg/s}$$

Pipes number 4 will also have to carry the heat required for emitters and pipework 1, 2 and 3.

total flow rate = 0.728 + 0.228

= 1 kg/s

From pipe-sizing Table 5.1, a 40 mm diameter steel pipe will carry 1.04 kg/s giving a pressure loss per metre run of 160 Pa and an EL factor of 1.8.
Check on estimated heat loss of 5 per cent (see Table 5.2).

19.5 m × 105 = 2047.5 watts

$$\text{percentage heat loss} = \frac{2047.5}{48\,000} \times \frac{100}{1}$$

heat loss = 4.3 per cent (approx.)

Pipes number 4 are therefore satisfactory for practical purposes;

Pump duty

Before the pressure required by the pump can be determined, it is necessary to find the effective length of each section of pipework in the system. The effective lengths of pipe are given by addition of the equivalent length due to the resistances of the fittings and the actual lengths.

Effective lengths (see Table 5.3)

Pipes number 1, 20 mm diameter with an actual length of 26 m.

bends	= 2 × 0.7 × 0.8 =	1.12 m
angle valves	= 2 × 0.5 × 0.8 =	0.80 m
tees	= 2 × 0.7 × 0.8 =	1.12 m
emitter	= 1 × 5.0 × 0.8 =	4.00 m
	total =	7.04 m

effective length = 26 + 7.04 = 33.04 m

Pipes number 2, 32 mm diameter with an actual length of 54 m.

bends	= 2 × 0.5 × 1.4 =	1.40 m
angle valves	= 2 × 0.5 × 1.4 =	1.40 m
tees	= 2 × 0.7 × 1.4 =	1.96 m
emitter	= 1 × 5.0 × 1.4 =	7.00 m
	total =	11.76 m

effective length 54 + 11.76 = 65.76 m

Pipes number 3, 32 mm diameter with an actual length of 26 m.

tees	= 2 × 0.7 × 1.5 =	2.1 m
angle valves	= 2 × 0.5 × 1.5 =	1.5 m
emitter	= 1 × 5.0 × 1.5 =	7.5 m
	total =	11.1 m

effective length = 26 + 11.1 = 37.1 m

Pipes number 4, 40 mm diameter with an actual length of 18.5 m.

bends	= 3 × 0.5 × 1.8 =	2.70 m
angle valves	= 2 × 0.5 × 1.8 =	1.80 m
tees	= 2 × 0.7 × 1.8 =	2.52 m
emitter	= 1 × 5.0 × 1.8 =	9.00 m
boiler	= 1 × 5.0 × 1.8 =	9.00 m
	total =	25.02 m

effective length = 18.5 + 25 = 43.50 m

Pump pressure

The pressure developed by the pump will have to overcome the frictional resistances through all the effective lengths of pipes to the index convector. A table may now be compiled (see Table 5.4).

Table 5.4 Total pressure loss on the system

Pipe number	Effective pipe length (m)	Pressure loss per metre run (Pa)	Pressure loss on section (Pa)
1	33.04	320.0	10 572.8
2	65.76	87.5	575.4
3	11.10	180.0	1 998.0
4	25.02	160.0	4 003.2
		Total pressure loss	17 149.4 Pa

A pump providing a pressure of 17.2 kPa (approx.) and giving a mass flow rate of 1 kg/s would be satisfactory.

Boiler power

The boiler will have to have sufficient power to supply heat required from the pipes and the convectors.

boiler power = 14.4 + 16.8 + 14.4 + 14.4

= 60 kW

A 10 per cent boiler margin may be added for pre-heating and to give a reserve of power for severe weather conditions.

Therefore the boiler power would be 66 kW.

Micro-pipe systems

Micro-pipe systems can be 'open' or 'closed' circuits. An open circuit is provided with an expansion and feed cistern with a cold feed and vent pipe, which are open to the atmosphere. A closed or sealed system dispenses with an expansion and feed cistern, and the open cold feed and vent pipes, and an expansion vessel is substituted in their place. This expansion vessel contains a diaphragm with nitrogen or air on one side and water from the heating system on the other. When the water is heated, the expansion is taken up by compressing the nitrogen or air. The Heating and Ventilating Contractors Association in its 'Guide to Good Practice' limits the flow-water temperature in open-circuit micro-pipe systems to 82 °C and in closed or sealed systems to 99 °C. Open circuits are usually operated with a temperature difference between the flow and return water of 11 °C, while sealed systems usually operate at a temperature difference between flow and return water of up to 20 °C. This higher temperature in the sealed systems increases the carrying capacity of the pipes.

The outside diameter of the pipes to the heat emitters may be 6 mm, 8 mm, or 10 mm, depending upon the heating load and the pump pressure.

Circuit design

The following factors are usually taken into account:

1. To prevent air-locking, the velocity of flow of water should be as high as possible; up to 1.5 m/s and above 0.3 m/s.
2. In determining the mass flow rate, an allowance is made for the heat loss from the pipework. This loss is generally taken as 5 per cent for the larger-diameter mains and 10 per cent for the micro-pipe, of the heat emission from the radiators each serve.

The following steps may be taken in finding the sizes of the heating pipes:

1. Calculate the heat requirements for each room as described in Chapter 1.
2. Make a sketch of the system showing the positions of the heat emitters, boiler and pipe runs. The micro-pipe system is always a two-pipe system.
3. Allocate a reference number to each section of pipework.
4. Decide on the temperature difference between the water in the flow and return pipes.
5. Estimate the mass flow rate through each circuit, allowing 5 or 10 per cent heat losses on the pipes.
6. Allowing a velocity of flow of water above 0.3 m/s and up to 1.5 m/s, find a suitable diameter pipe from Tables 5.5 and 5.6.
7. Measure the actual lengths of the circuits and to these add an allowance for frictional losses due to fittings to obtain the effective pipe length.
8. Find the 'index run' and determine the pressure drop along it in order to ascertain the pump pressure.

An example will show how these steps are carried out.

Example 5.3 *Figure 5.3 shows a micro-pipe system for a three-bedroomed house. If an 'open' type of system is to be used with flow and return water temperatures of 80 °C and 69 °C respectively, find the approximate diameters of the pipes and the pump duty.*

Pipes number 1

$$\begin{aligned}
\text{emission from radiator} &= 0.900 \text{ kW}\\
\text{emission from pipework} &= 0.090 \text{ kW}\\
\text{(10 per cent of 0.9 kW)}&\\
\text{total} &= 0.990 \text{ kW}
\end{aligned}$$

$$\text{flow rate} = \frac{0.99}{4.2 \times (80-69)}$$

$$= 0.0214 \text{ kg/s}$$

From the pipe-sizing table (Table 5.5), an 8 mm outside diameter tube will carry 0.0217 kg/s giving a pressure loss per metre run of 902 Pa.

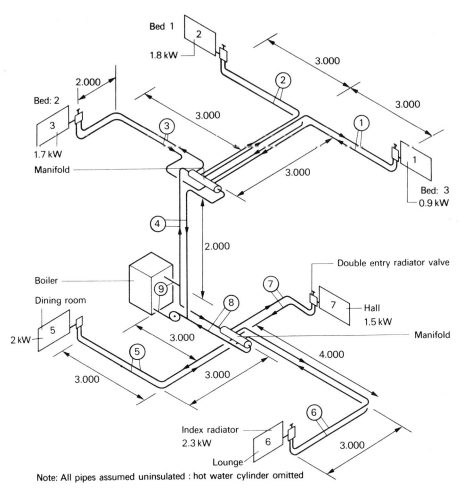

Fig. 5.3   Micro-pipe system

Pipes number 2

$$\begin{aligned}
\text{emission from radiator} &= 1.800 \text{ kW}\\
\text{emission from pipework} &= 0.180 \text{ kW}\\
\text{(10 per cent of 1.8 kW)}&\\
\text{total} &= 1.980 \text{ kW}
\end{aligned}$$

$$\text{flow rate} = \frac{1.98}{4.2 \times 11}$$

$$= 0.043 \text{ kg/s}$$

From the pipe-sizing Table 5.5, a 10 mm outside diameter tube will carry 0.0433 kg/s giving a pressure loss per metre run of 902 Pa.

Pipes number 3

$$\text{emission from radiator} = 1.700 \text{ kW}$$

62

Table 5.5 Table for the flow of water at 82°C in small-diameter copper pipes (outside diameter)

Diameter 6 mm	Diameter 8 mm	Diameter 10 mm	Velocity	Pressure drop (Pa per m run of pipe)
(kg/s)	(kg/s)	(kg/s)	(m/s)	
0.005 6	0.013 8	0.027 7		410.0
0.006 3	0.015 3	0.030 6		492.0
0.006 8	0.016 8	0.033 5		574.0
0.007 4	0.018 2	0.036 2		656.0
0.007 9	0.019 5	0.038 8	0.6	738.0
0.008 4	0.020 6	0.041 0		820.0
0.008 8	0.021 7	0.043 3		902.0
0.009 3	0.022 9	0.045 6		984.0
0.009 6	0.024 0	0.047 6		1066.0
0.010 2	0.025 1	0.049 8		1148.0
0.010 5	0.026 1	0.051 6	0.9	1230.0
0.010 9	0.027 1	0.053 5		1312.0
0.011 3	0.027 9	0.055 4		1394.0
0.011 6	0.029 0	0.057 4		1476.0
0.012 1	0.029 8	0.058 7		1558.0
0.012 4	0.030 6	0.060 8		1640.0
0.013 3	0.032 8	0.065 1		1845.0
0.014 2	0.034 9	0.069 0		2050.0
0.015 0	0.036 8	0.072 9	1.2	2255.0

$$\text{emission from pipework} = 0.170 \text{ kW}$$
(10 per cent of 1.7 kW)

$$\text{total} = 1.870 \text{ kW}$$

$$\text{flow rate} = \frac{1.870}{4.2 \times 11}$$

$$= 0.040 \text{ kg/s}$$

From the pipe-sizing Table 5.5, a 10 mm outside diameter tube will carry 0.041 kg/s giving a pressure loss per metre run of 820 Pa.

Pipes number 4

These pipes will have to carry the heat loads required for pipes 1, 2 and 3 plus the heat emission from the pipes number 4.

$$\begin{aligned}\text{heat load pipes 1} &= 0.990 \text{ kW}\\ \text{heat load pipes 2} &= 1.980 \text{ kW}\\ \text{heat load pipes 3} &= 1.870 \text{ kW}\\ &= 4.840 \text{ kW}\end{aligned}$$

$$\text{emission from pipes number 4} = 0.242$$
(5 per cent of 4.84 kW)

$$\text{total} = 5.082 \text{ kW}$$

$$\text{flow rate} = \frac{5.082}{4.2 \times 11}$$

$$= 0.110 \text{ kg/s}$$

From pipe-sizing Table 5.6, a 22 mm outside diameter tube will carry 0.111 kg/s giving a pressure loss per metre run of pipe of 80 Pa.

Pipes number 5

$$\begin{aligned}\text{emission from radiator} &= 2.00 \text{ kW}\\ \text{emission from pipes} &= 0.20 \text{ kW}\end{aligned}$$
(10 per cent of 2 kW)

$$\text{total} = 2.20 \text{ kW}$$

$$\text{flow rate} = \frac{22}{4.2 \times 11}$$

$$= 0.048 \text{ kg/s}$$

From the pipe-sizing Table 5.5, a 10 mm outside diameter tube will carry 0.0498 kg/s giving a pressure loss per metre run of 1148 Pa.

Pipes number 6

$$\begin{aligned}\text{emission from radiator} &= 2.300 \text{ kW}\\ \text{emission from pipes} &= 0.230 \text{ kW}\end{aligned}$$
(10 per cent of 2.3 kW)

$$\text{total} = 2.530 \text{ kW}$$

$$\text{flow rate} = \frac{2.530}{4.2 \times 11}$$

$$= 0.054 \text{ kg/s}$$

From the pipe-sizing Table 5.5, a 10 mm outside diameter tube will carry 0.0554 kg/s giving a pressure loss per metre run of 1394 Pa.

Pipes number 7

$$\begin{aligned}\text{emission from radiator} &= 1.500 \text{ kW}\\ \text{emission from pipes} &= 0.150 \text{ kW}\end{aligned}$$
(10 per cent of 1.5 kW)

$$\text{total} = 1.650 \text{ kW}$$

$$\text{flow rate} = \frac{1.650}{4.2 \times 11}$$

$$= 0.036 \text{ kg/s}$$

From the pipe-sizing Table 5.5, a 10 mm outside diameter tube will carry 0.0362 kg/s giving a pressure loss per metre run of 656 Pa.

Pipes number 8

These will have to carry the heat loads for pipes 5, 6 and 7 plus the heat emission from pipes number 8.

heat load pipes 5 = 2.200 kW
heat load pipes 6 = 2.530 kW
heat load pipes 7 = 1.650 kW
total = 6.380 kW

Pipes number 8 are very much shorter in length than pipes 5, 6 and 7, therefore the estimated heat emission from them will be taken as 5 per cent of the above total.

Heat load to be carried by pipes number 8 is therefore:

6.38 kW + 5 per cent of 6.38

$$= 6.38 + 0.319$$
$$= 6.699 \text{ kW}$$

$$\text{flow rate} = \frac{6.699}{4.2 \times 11}$$

$$= 0.145 \text{ kg/s}$$

From the pipe-sizing Table 5.6, a 22 mm outside diameter tube will carry 0.152 kg/s giving a pressure loss per metre run of 140 Pa.

Table 5.6 Table for the flow of water at 80°C in copper pipes (outside diameter). Block table to BS 2871-1-X

Diameter 12 mm	Diameter 15 mm	Diameter 22 mm	Diameter 28 mm	Diameter 35 mm	Velocity	Pressure drop (Pa per m run of pipe)
(kg/s)	(kg/s)	(kg/s)	(kg/s)	(kg/s)	(m/s)	
0.016	0.030	0.089	0.181	0.181	0.3	55.0
0.017	0.032	0.094	0.191	0.344		60.0
0.017	0.033	0.098	0.199	0.360		65.0
0.018	0.035	0.103	0.208	0.376		70.0
0.019	0.036	0.107	0.216	0.390		75.0
0.020	0.037	0.111	0.224	0.405	0.5	80.0
0.021	0.040	0.119	0.240	0.433		90.0
0.022	0.043	0.126	0.254	0.459		100.0
0.025	0.047	1.140	0.282	0.508		120.0
0.027	0.052	0.152	0.308	0.554		140.0
0.030	0.056	0.164	0.332	0.597		160.0
0.032	0.060	0.176	0.354	0.638		180.0
0.034	0.064	0.186	0.376	0.676		200.0
0.038	0.071	0.207	0.416	0.749	1.0	240.0

Pipes number 9

These will have to carry the heat loads for pipes 1, 2, 3, 5, 6, and 7 plus the heat emission from pipes number 4 and 8.

heat load pipes 1 = 0.990 kW
heat load pipes 2 = 1.980 kW

heat load pipes 3 = 1.870 kW
heat load pipes 4 = 0.242 kW
heat load pipes 5 = 2.200 kW
heat load pipes 6 = 2.530 kW
heat load pipes 7 = 1.650 kW
heat load pipes 8 = 0.319 kW
11.781 kW

Pipes number 9 are again very short and an estimated heat loss from them of 5 per cent of the above total will be adequate.

heat load to be carried
by pipes number 9 = 11.781 + 5 per cent of 11.781
= 11.781 + 0.589
= 12.370 kW

$$\text{flow rate} = \frac{12.37}{4.2 \times 11}$$

$$= 0.267 \text{ kg/s}$$

From the pipe-sizing Table 5.6, a 28 mm outside diameter tube will carry 0.282 kg/s giving a pressure loss per metre run of 120 Pa. A check may be made on the flow rate through pipes number 9 as follows:

flow rate through
pipes number 4 = 0.110 kg/s
flow rate through
pipes number 8 = 0.145 kg/s
= 0.255 kg/s

To this flow rate an allowance must be made for the heat emission from pipes number 9.

heat loss from pipes number 9 = 0.589 kW
(5 per cent of 11.781 kW)

$$\text{flow rate} = \frac{0.589}{4.2 \times 11}$$

$$= 0.0127 \text{ kg/s}$$

total flow rate through
pipes number 9 = 0.255 + 0.0127
= 0.267 kg/s

This corresponds with the flow rate previously calculated.

Boiler power

The boiler power must be sufficient to supply heat for the radiators, emission from all the pipes, and to heat the water in the storage calorifier.

power required for radiators
and pipes = 12.370 kW
power required for hot water
supply = 3.000 kW
total = 15.370 kW

A 10 per cent boiler margin may be added for pre-heating and to give a reserve of power for severe weather conditions.

Therefore the boiler power would be:

15.37 kW + 10 per cent = 16.907 kW

A 17 kW boiler would be suitable.

Pump duty

The effective length of the index circuit must first be found by adding an allowance for fittings of 30 per cent on the mains and 10 per cent on the micro-pipe circuit. The index circuit will be the one supplying radiator number 6.

effective length of 10 mm tube

$$24 \text{ m} + 10 \text{ per cent} = 26.4 \text{ m}$$
$$\text{resistance} = 26.4 \times 1.394 \text{ kPa}$$
$$= 36.96 \text{ kPa}$$

effective length of 22 mm and 28 mm tube

$$7 \text{ m} + 30 \text{ per cent} = 9.1 \text{ m}$$
$$\text{resistance} = 9.1 \times 0.140 \text{ kPa}$$
$$= 1.274 \text{ kPa}$$
$$\text{total resistance} = 36.96 + 1.274$$
$$= 38.234 \text{ kPa}$$

A pump providing a pressure of 40 kPa and giving a mass flow rate of 0.3 kg/s would be satisfactory.

Pipe-sizing for gravity circulation

Gravity-circulation systems are not nowadays used for space-heating systems for anything other than a very small building. The system is, however, still used extensively for the primary circulation between the boiler and calorifier in a hot-water supply installation.

Circulating pressure

The pressure required to ensure circulation in gravity systems is termed 'circulation pressure' and is expressed in pascals per metre of circulating height, i.e. the height measured between the centres of the flow and return connections to the boiler and radiator respectively (see Fig. 5.4).

The circulating pressure per metre of height is found from the following formula:

$$\text{circulating pressure} = 9.81 (\rho_2 - \rho_1)$$
(Pa per metre height)

where ρ_2 = density of return water/kg/m^3)

ρ_1 = density of flow water (kg/m^3)

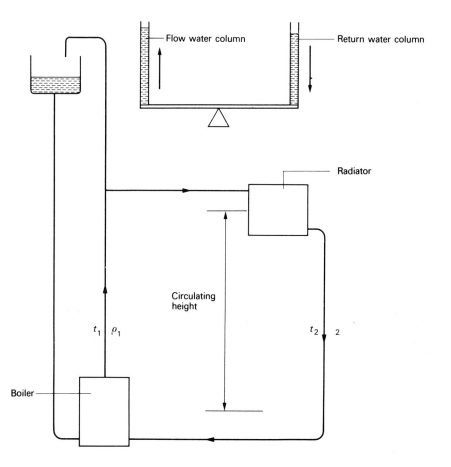

Fig. 5.4 Circulation by natural convection

Example 5.4 *Calculate the circulating pressure in pascals when the flow and return water are 82.2°C and 60°C respectively and the circulating height is 3m. See Table 5.7 for density of water at various temperatures.*

$$\text{total circulating pressure} = H \times 9.81 (\rho_2 - \rho_1)$$
$$= 3 \times 9.81 (983.24 - 970.43)$$
$$= 3 \times 9.81 \times 12.81$$
$$= 376.998 \text{ Pa}$$

This circulating pressure must be sufficient to overcome the pressure due to frictional resistance of the pipe and fittings.

Circulating pressure per metre run

In order to find a suitable diameter of pipe from Table 5.8, it is necessary to find the circulating pressure per metre run of pipe. The run of pipe is the actual length plus the equivalent length due to the resistances of fittings.

Table 5.7 Density of water at various temperatures

Temperature (°C)	Density (kg/m³)
43.3	990.93
48.9	988.56
54.4	986.05
60.0	983.24
65.6	980.29
71.0	977.13
76.7	973.88
82.2	970.43
87.8	966.82
93.3	963.07

Table 5.8 Table for the flow of water at 75°C in black steel pipes. Heavy-grade pipes to BS 1387

Diameter 20 mm		Diameter 25 mm		Diameter 32 mm			Pressure drop (Pa per m run of pipe)
M	EL	M	EL	M	EL	V	
0.020	0.5	0.037	0.7	0.082	1.0		4.0
0.021	0.5	0.039	0.7	0.087	1.0		4.5
0.022	0.5	0.042	0.7	0.093	1.0		5.0
0.023	0.5	0.044	0.7	0.098	1.0		5.5
0.025	0.5	0.046	0.7	0.103	1.1		6.0
0.026	0.5	0.048	0.7	0.107	1.1		6.5
0.027	0.5	0.050	0.7	0.112	1.1		7.0
0.028	0.5	0.052	0.7	0.116	1.1		7.5
0.029	0.5	0.054	0.7	0.120	1.1		8.0
0.030	0.5	0.056	0.7	0.125	1.1		8.5
0.031	0.5	0.058	0.7	0.129	1.1		9.0
0.032	0.5	0.060	0.7	0.133	1.1		9.5
0.033	0.5	0.062	0.7	0.136	1.1	0.15	10.0
0.037	0.5	0.070	0.8	0.154	1.1		12.0
0.042	0.6	0.077	0.8	0.171	1.2		15.0
0.045	0.6	0.084	0.8	0.186	1.2		17.5
0.049	0.6	0.091	0.8	0.200	1.2		20.0
0.052	0.6	0.097	0.8	0.214	1.2		22.5
0.055	0.6	0.103	0.8	0.226	1.2	.030	25.0

V = velocity of flow m/s
M = mass rate of flow kg/s
EL = equivalent length factor

Example 5.5 (see Fig. 5.5). *Find by use of Table 5.8 the diameter of the primary flow and return pipes for a hot-water cylinder holding a mass of 136 kg of water which is to be raised in temperature from 10°C to 71°C in 2 hours. The temperatures of the flow and return pipes are to be 71°C and 60°C respectively and the circulating height 1.5 m. An allowance of 20 per cent for heat losses on the pipes and a frictional loss of 10 per cent on the actual length of pipes may be used.*

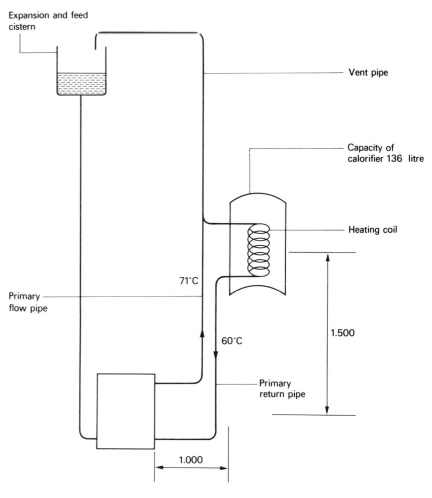

Fig. 5.5 Primary circuit

$$\text{power} = \frac{\text{s.h.c.} \times \text{kg} \times (\text{temperature rise °C}) \times 100}{3600 \times 2 \times \text{efficiency}}$$

$$= \frac{4.2 \times 136 \times (71-10) \times 100}{3600 \times 2 \times 80}$$

$$= 6.049 \text{ kW}$$

$$\text{flow rate} = \frac{6.049}{4.2 \times 11}$$

$$= 0.131 \text{ kg/s}$$

$$\text{total circulating pressure} = H \times 9.81 \,(\rho_2 - \rho_1)$$

$$= 1.5 \times 9.81 \times (983.24 - 977.13)$$

$$= 1.5 \times 9.81 \times 6.11$$

$$= 90 \text{ Pa}$$

$$\text{effective length of pipe} = 5 + 10 \text{ per cent}$$

$$= 5.5 \text{ m}$$

$$\text{circulating pressure available per metre run} = \frac{\text{total circulating pressure}}{\text{effective length}}$$

$$= \frac{90}{5.5}$$

$$= 16.36 \text{ Pa}$$

From the pipe-sizing Table 5.8, a 32 mm diameter steel pipe will carry 0.133 kg/s giving a pressure loss of 9.5 Pa per metre run. Since the flow rate for this pipe is greater than required and the frictional loss is less, the pipe will be satisfactory.

Sizing of secondary circuits

The secondary flow pipe will have to deliver a certain rate of flow to the various sanitary fittings, while the secondary return will have to carry sufficient water to compensate for the heat losses from the pipes and any heat emitted from towel rails, etc. The secondary return may therefore be determined as previously described for the pipe-sizing for heating.

The sizing of the secondary flow pipe can be carried out from charts or tables or by the Box formula described in Chapter 3, using the discharge rates from the sanitary fittings as listed in Table 5.9, and allowing a diversity factor.

Table 5.9 Recommended minimum rate of flow at various appliances

Type of appliance	Rate of flow (litre/s)
WC flushing cistern	0.12
Wash basin	0.15
Wash basin with spray taps	0.04
Bath (private)	0.30
Bath (public)	0.60
Shower (with nozzle)	0.12
Sink with 13 mm taps	0.20
Sink with 19 mm taps	0.30
Sink with 25 mm taps	0.60

Index circuit (see Fig. 5.6)

In a gravity-heating system the index circuit may be defined as the circuit in which the ratio of the circuit length to the circuit circulating height is the minimum.

If the temperature drop across each circuit in Fig. 5.6 is the same, and since circuit 1 has twice the circulating height of circuit 2, it will have double the circulating pressure of circuit 2. Since circuit 2 has a greater length than circuit 1, it follows that circuit 2 is the index circuit.

In order to find the diameters of pipes for the system, the circulating pressure per metre of travel will be based on circuit 2. The method of pipe-sizing will then be carried out as described for the primary flow and return pipes in a hot-water supply system.

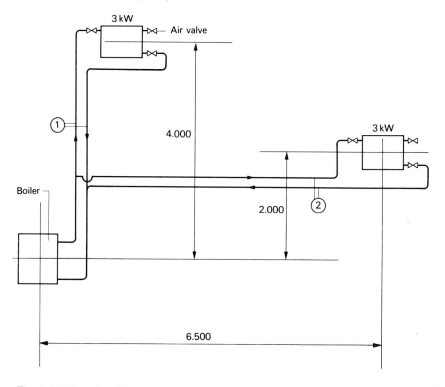

Fig. 5.6 Index circuit

Questions

1. Figure 5.7 shows a single-pipe circuit, assuming each radiator has a heat emission of 5 kW and the temperature drop across the system is 10 °C. Find by use of Table 5.10 the approximate diameter of a copper main pipe. A velocity of flow of water is to be 0.5 m/s. Determine the pump duty for the circuit allowing an emission from the pipes 20 per cent of the total from the radiators.

67

Answers: 51 mm diameter; pump duty – pressure 3 kPa; discharge 0.86 kg/s
(0.9 kg/s approx.)

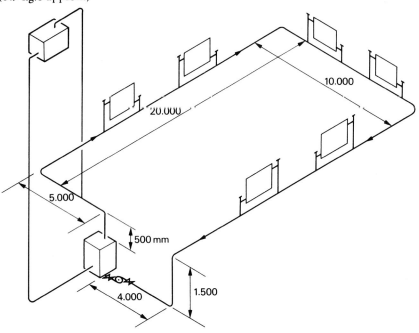

Fig. 5.7

Table 5.10 Table for the flow of water at 75°C in copper tubes BS 659 Light Gauge

Diameter 51 mm		Diameter 63 mm		Diameter 76 mm			Pressure drop (Pa per m run of pipe)
M	EL	M	EL	M	EL	V	
0.428	2.2	0.774	2.9	1.26	3.8	0.30	10.0
0.486	2.2	0.878	3.0	1.42	3.8		12.5
0.538	2.3	0.972	3.1	1.58	3.9		15.0
0.587	2.3	1.06	3.1	1.72	4.0		17.5
0.633	2.4	1.14	3.2	1.85	4.1		20.0
0.676	2.4	1.22	3.2	1.98	4.1		22.5
0.714	2.4	1.29	3.3	2.10	4.2		25.0
0.757	2.5	1.37	3.3	2.21	4.2		27.5
0.795	2.5	1.43	3.3	2.32	4.3		30.0
0.838	2.5	1.50	3.4	2.42	4.3		32.5
0.867	2.5	1.56	3.4	2.53	4.3		35.0
0.907	2.5	1.62	3.4	2.63	4.4	0.50	37.5

V = velocity of flow m/s
M = mass rate of flow kg/s
EL = equivalent length factor

68

2. Figure 5.8 shows a two-pipe circuit assuming each radiator has a heat emission of 8 kW and the temperature drop across the system is 10°C. Find by use of Table 5.11 the approximate diameters of the steel pipes. Determine the pump duty for the circuit allowing an emission from each section of the pipework 5 per cent of the total from the radiators.

Answers: pipe number 1, 20 mm diameter; pipe number 2, 25 mm diameter; pump duty – pressure 5 kPa; discharge 0.3 kg/s.

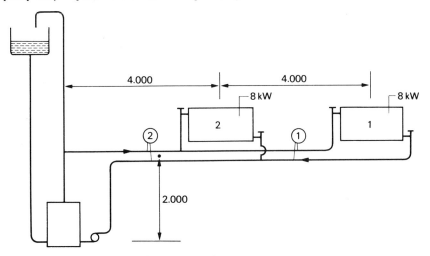

Fig. 5.8

Table 5.11 Table for the flow of water at 75°C in black steel pipes. Heavy-grade pipes to BS 1387

Diameter 15 mm		Diameter 20 mm		Diameter 25 mm			Pressure drop (Pa per m run of pipe)
M	EL	M	EL	M	EL	V	
0.011	0.3	0.026	0.5	0.048	0.7	0.15	6.5
0.011	0.3	0.027	0.5	0.050	0.7	0.15	7.0
0.012	0.3	0.028	0.5	0.052	0.7	0.15	7.5
0.012	0.3	0.029	0.5	0.054	0.7	0.15	8.0
0.013	0.3	0.030	0.5	0.056	0.7	0.15	8.5
0.013	0.3	0.031	0.5	0.058	0.7	0.15	9.0
0.014	0.3	0.032	0.5	0.060	0.7	0.15	9.5
0.014	0.3	0.033	0.5	0.062	0.7	0.15	10.0
0.016	0.3	0.037	0.5	0.070	0.8	0.15	12.5
0.018	0.4	0.042	0.6	0.077	0.8	0.15	15.0
0.019	0.4	0.045	0.6	0.084	0.8	0.15	17.5
0.021	0.4	0.049	0.6	0.091	0.8	0.15	20.0

V = velocity of flow m/s
M = mass rate of flow kg/s
EL = equivalent length factor

3. Define the following terms: (*a*) index circuit; (*b*) index radiator; (*c*) circulating pressure; (*d*) circulating height.

4. Figure 5.9 shows a part of a micro-pipe system which represents the index circuit. If an 'open' type of system is to be used with flow and return water temperatures of 80 °C and 69 °C respectively, find the diameters of the micro-pipes and the total pressure drop in pascals through them.

Answers: 10 mm O.D. copper tube; pressure drop 24 805 Pa or 24.8 kPa.

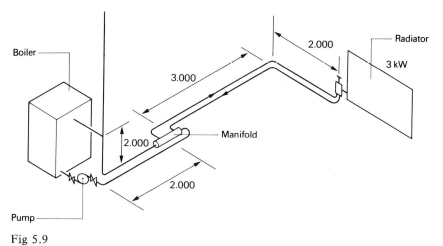

Fig 5.9

5. Calculate the total circulating pressure for a gravity heating system when the circulating height is 4 m and the flow and return waters are 82.2 °C and 60 °C respectively.

Answer: 502.3 Pa

6. Find by the use of Table 5.8 suitable diameter flow and return pipes for a 136 litre hot-water cylinder. Use the following factors:

(*a*) heat recovery period = 3 hours
(*b*) circulating height = 1.2 m
(*c*) actual length of pipes = 4.4 m
(*d*) temperature flow of water = 71 °C
(*e*) temperature of return water = 60 °C
(*f*) temperature rise of water in the cylinder = 65 °C
(*g*) heat loss in pipes = 20 per cent of power required to heat water
(*h*) frictional loss on pipes = 10 per cent of actual length

Answer: 32 mm diameter steel pipe

Chapter 6

Gas heating, flues and electric heating

Gas space heating

Boilers

Gas-fired, wall-hung boilers are popular because of saving in floor space. They can be obtained with either a conventional or a balanced flue. The heat exchanger of the boiler consists of a stainless steel or copper tube which holds approximately 0.7 litre of water, and this low water content permits a rapid thermal response. A typical wall-hung boiler has a heat output of 15 kW which is sufficient to heat domestic hot water and up to 20 m² of radiator surface, including the pipe runs.

Hearth-mounted boilers have heat outputs ranging from 8.8 kW for a small house to 1760 kW for a large building. The smaller boilers can be obtained with either a conventional or a balanced flue, but larger boilers are normally provided with a conventional flue. Figure 6.1 shows a small gas-fired, hearth-mounted boiler including the controls. These have been shown outside the boiler for clarity, but they are usually fitted inside the boiler casing.

Position

The boiler can be placed on the rooftop or in any other convenient position since there is no storage of fuel and removal of ashes are no problem. This is of great value since ground-floor and basement space is exceptionally valuable. A shorter conventional flue is also possible which saves on construction costs and

space. There is also less pressure on the boiler. All types of boilers are fully automatic in operation and incorporate thermostatic and clock control, a flame-failure device which ensures a 100 per cent gas cut-off in the event of pilot failure, and a constant pressure governor.

Efficiencies are high — of the order of 75–80 per cent — and space requirements per unit of heat output is slightly less than for solid fuel and oil-fired boilers.

Construction

Sectional cast-iron boilers are available with heat outputs of between 8.8 and 440 kW. It is possible to erect the boilers in confined spaces and a single section can be readily replaced.

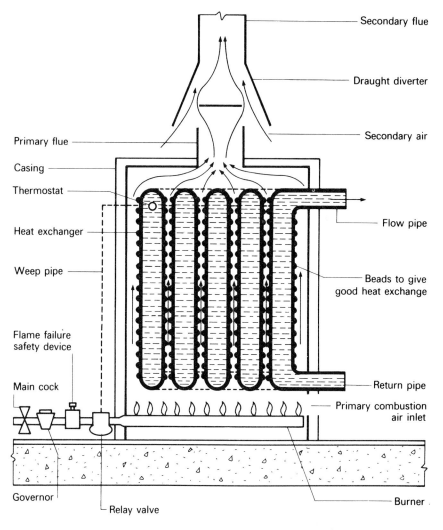

Fig.6.1 Floor-mounted gas boiler

Welded plate steel boilers are normally fabricated in one piece, which reduces site work. Heat outputs of up to 180 kW are available. Welded tubular-steel boilers are available for larger installations, with heat outputs of up to 1760 kW. Steel boilers can withstand higher pressures than cast-iron boilers.

Fires

These are usually designed to provide both radiant and convective heating to the room; in addition they can incorporate a back boiler at the rear which can be easily fitted into a standard fireplace opening.

A typical fire has a heat output of 3 kW, and a gas-fired back boiler a heat output of 12 kW, which is sufficient to heat the normal domestic hot water and up to 18 m² of radiator surface, including the pipe runs. If domestic hot water is not required a back boiler will heat up to 20 m² of radiator surface, including the pipe runs. Fires can be either hearth- or wall-mounted. Figure 6.2 shows a section through a hearth-mounted gas fire providing both radiant and convective space heating.

Convector heaters

These are usually room-sealed combustion units, incorporating a balanced flue. The heat output from the convector is almost all by convection although some radiant heat is given off from the hot casing, usually about 10 per cent. They provide a simple, flexible space heating system and are usually cheaper to install than a boiler and radiator system. Figure 6.3 shows a natural convector and Fig. 6.4 a fan-assisted convector which gives a rapid thermal response. Heaters are available with heat outputs of between 5 kW and 7 kW.

Warm-air units (see Fig. 6.5)

These are free-standing, self-contained packaged units which combine controlled heating, ventilation and air movement. They are provided with a silent fan and built-in noise attenuators which make them suitable for the space heating of houses, schools, churches and libraries. The warm air may be ducted from the unit to the various rooms, as shown in Fig. 6.6. The ducts must be well insulated to prevent heat loss. Units are available having heat outputs from 6 kW to 50 kW.

A warm-air space heating system provides a good distribution of heat and the air movement gives the required feeling of freshness. The system can be used in hot weather to circulate the air in the rooms. However, criticism of the system is that there is no radiant heating and a gas fire may be required to provide this, especially in the lounge.

Unit heaters for overhead use

These are suitable for factories, churches and assembly halls and two types are available:

1. Indirect, which have a flue to carry off the products of combustion and have a heat output of up to 230 kW at 75 per cent efficiency.
2. Direct or flueless heater, in which the products of combustion are circulated together with the heated air and have a heat output of up to 41 kW at 90 per cent efficiency.

Figure 6.7 shows a direct overhead unit heater installed in a factory or assembly hall.

Radiant heaters

Thermal comfort may be provided for the occupants of a room at comparatively low capital and running costs, by use of radiant heaters mounted either on the wall or ceiling. They can be installed in factories, churches and assembly halls and are used as an alternative to the convector heaters for heating these buildings. Thermal comfort conditions can be obtained quicker with radiant heaters than with convector heaters and the pre-heating time and fuel costs are therefore reduced to a minimum. Figure 6.8 shows a wall- or ceiling-mounted radiant heater.

Another type of radiant heater consists of a 64 mm bore steel U tube into which a gas burner is fixed. A silent fan draws the products of combustion through the tube which is heated before the gases are discharged to atmosphere. An insulated polished reflector is fitted above the tube to ensure maximum radiation of heat.

Flues for gas appliances

Gas flues are simpler and cheaper to construct than flues for solid fuel or oil and they do not require periodic cleaning. Certain gas appliances such as space heaters and cookers do not require a flue and are permitted to discharge their products of combustion into the room in which they are installed. All other appliances require a flue and the local Gas Board should be consulted on this matter. Gas appliances have to be able to function without a flue, and the flue is therefore only required to discharge the products of combustion to the atmosphere and not to create a draught to aid combustion.

The products of combustion of gas are clean and comply with the Clean Air Act. The type of flue for a specific project depends upon several factors which include the height and type of construction of the building, the type and siting of the appliance, wind conditions and current building regulations. These factors are interacting and must therefore be considered as a whole. The products of combustion from gas appliances contain water vapour, and the placing and design of the flue should either prevent condensation or remove any water resulting from it.

Terms used

Draught diverter. A device designed for preventing downdraught or static conditions in the secondary flue of an appliance from interfering with the combustion of gas within the appliance. It also prevents excessive draught conditions by allowing the air in the room to mix with the products of combustion in the secondary flue and then cool down these hot gases. Figure 6.9 shows the operation of a draught diverter.

Duct. A tube or casing used for the passage of the products of combustion or air.

Excess air. Air in excess of that theoretically required for complete combustion of gas.

Flue. A tube or casing used for the passage of the products of combustion.

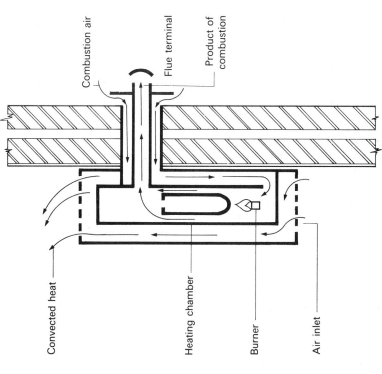

Combustion air

Flue terminal

Product of combustion

Convected heat

Heating chamber

Burner

Air inlet

Fig. 6.3 Natural convector

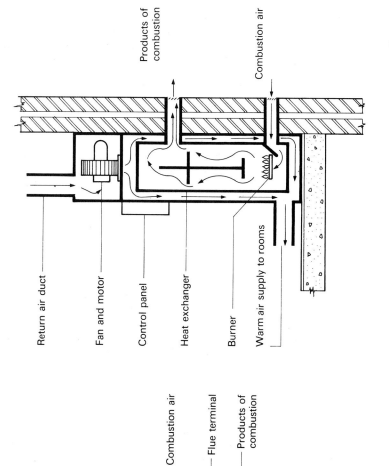

Products of combustion

Combustion air

Return air duct

Fan and motor

Control panel

Heat exchanger

Burner

Warm air supply to rooms

Fig. 6.5 Gas warm-air unit

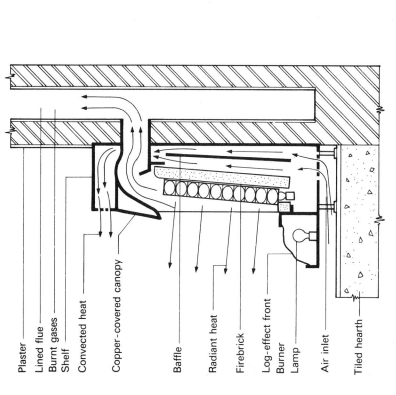

Plaster

Lined flue

Burnt gases

Shelf

Convected heat

Copper-covered canopy

Baffle

Radiant heat

Firebrick

Log-effect front

Burner

Lamp

Air inlet

Tiled hearth

Fig. 6.2 Built-in-type gas fire

Combustion air

Flue terminal

Products of combustion

Convected heat

Heating chamber

Burner

Fan

Air inlet

Fig. 6.4 Fan-assisted convector

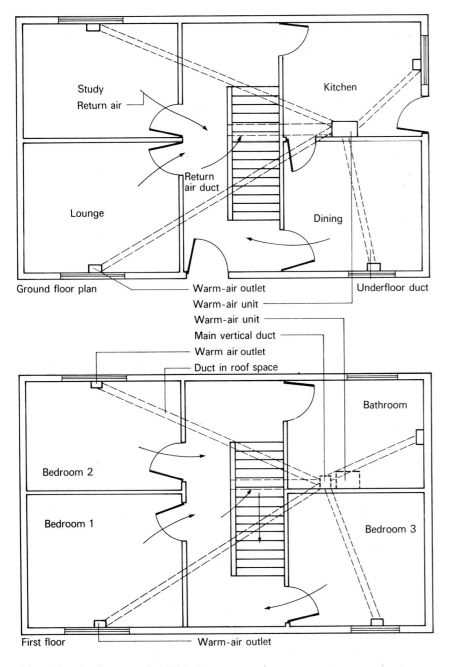

Ground floor plan

Study
Return air

Return
air duct

Kitchen

Lounge

Dining

Warm-air outlet
Underfloor duct

Warm-air unit

Warm-air unit
Main vertical duct
Warm air outlet
Duct in roof space

Bedroom 2

Bathroom

Bedroom 1

Bedroom 3

First floor
Warm-air outlet

Note: When the doors are closed the air
returns through grills in the doors

Fig. 6.6 Warm-air system

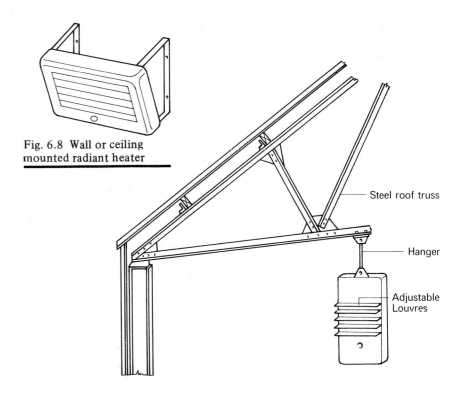

Fig. 6.8 Wall or ceiling
mounted radiant heater

Steel roof truss

Hanger

Adjustable
Louvres

Fig. 6.7 Direct-unit heater

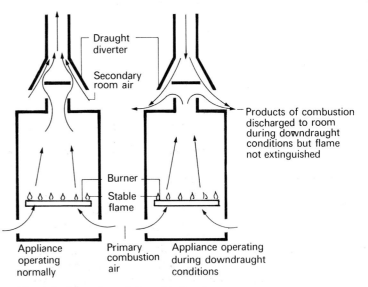

Draught
diverter

Secondary
room air

Products of combustion
discharged to room
during downdraught
conditions but flame
not extinguished

Burner

Stable
flame

Appliance
operating
normally

Primary
combustion
air

Appliance operating
during downdraught
conditions

Fig. 6.9 Operation of draught diverter

Primary flue. A length of flue prior to the draught diverter (see Fig. 6.10)

Secondary flue. This is the flue proper and is the flue between the draught diverter and the terminal (see Fig. 6.10).

Main flue. A flue used for carrying the products of combustion from two or more appliances (see Fig. 6.11).

Subsidiary flue. A flue connecting an appliance with the main flue (see Fig. 6.11).

Individual flue. A flue serving only one appliance.

Branched or shunt flue system. A flue system comprising a main flue into which the products of combustion from two or more appliances discharge, each by way of a vertical subsidiary flue (see Fig. 6.12).

Common flue system. A flue system taking the products of combustion from more than one appliance. It includes a branched flue system.

Room-sealed appliance. An appliance having the air inlet and flue outlets (except for the purpose of lighting) *sealed* from the room in which the appliance is installed. It includes a drying cabinet having an access door, with means of automatically closing the air inlet and flue outlet when the door is opened.

Balanced-flue appliance. An appliance designed to draw in combustion air from a point immediately adjacent to where it discharges its products of combustion. The inlet and outlet points are incorporated in a windproof terminal, which is sited outside the room in which the appliance is fitted (see Fig. 6.13 which shows the principles of operation of a balanced-flue appliance).

SE duct. A duct rising vertically which is open at the bottom and top, serving to bring combustion air to and take the products of combustion from room-sealed appliances to the external air (see Fig. 6.14).

U duct. A duct in the form of a U to one limb of which are fitted room-sealed appliances, while the other limb provides combustion air (see Fig. 6.15).

Terminal. A device fitted at the termination of the flue, designed to allow free passage of the products of combustion, to minimise downdraught and to prevent the entry of foreign matter which might cause restriction of the flue. Figure 6.16 shows various types of terminals.

Venting. The removal of the products of combustion from an appliance.

Principles of design of flues

A gas flue in a building may serve two purposes namely:

1. To remove the products of combustion from an appliance.
2. To assist in the ventilation of the room in which the appliance is installed.

The force causing the movement of the gases inside a flue is usually due to the difference in density between the hot gases in the flue and the cooler air inside the room. This force is small and it is therefore essential that bends and terminals used in the construction of the flue should offer low resistance to the flow of gases. The flue should also terminate in such a position in the open air that the effects of wind pressure will aid the updraught and not act adversely.

Note: Some gas flues use a fan to extract the products of combustion from the appliance, and therefore the force causing the movement of the gases in the flue is not due solely to the differences in density between the hot gases and the cooler air inside the room. However, this natural convection will reduce the

power required for the fan. The size of a flue for a gas appliance is dependent upon the kilowatt rating of the appliance and the ventilation standard of the room in which the appliance is installed.

Materials for flues

1. Asbestos-cement pipes and fittings suitable for internal and external work and should conform to BSS 567 or BSS 835.
2. Steel or cast-iron pipes protected by a good-quality vitreous enamel.
3. Sheet aluminium (except where temperatures and condensation is liable to be excessive, e.g. a water-heater flue in a cold position).
4. Sheet copper protected by chromium plating after fabricating.
5. Stainless steel.
6. Precast concrete flue blocks (see Fig. 6.17) made of acid resisting cement and pointed by acid resisting cement mortar.
7. Brick flue, lined with precast concrete made of acid resisting cement and pointed with acid resisting cement mortar, or glazed stoneware pipes pointed with high-acid resisting mortar.

Planning of flues

All necessary information with regards to the installation of gas flues, should be made at the early design stage of the building. The local Gas Board should be consulted before any drawings and specifications are made and also during the execution of the work. The drawings should show the proposed positions of appliances and full details of the routes of all flues, from the point of connection to the appliances to the position of termination.

Flue installations must conform with the Buildings Regulations, 1972. The following points must be considered when planning a flue:

1. The flue should rise progressively towards its terminal.
2. The primary flue should be as short as possible.
3. The secondary flue should be of adequate height and kept inside the building so far as is possible.
4. In many appliances the draught diverter is an integral part of the appliance and permits easy disconnection of the flue from the appliance. Where an appliance is not fitted with a draught diverter, a disconnecting device should be fitted as near as is practicable to the appliance.
5. Routes which expose the flue to rapid cooling should be avoided, or thermal insulation afforded to the flue external surface.
6. Horizontal flues and fittings having sharp angles must be avoided.
7. The joints of asbestos-cement pipes and fittings should be made by caulking with slag wool for about 25 per cent of the depth of the socket. The remaining 75 per cent of the socket should be filled with a good-quality fire cement and the outside edge neatly chamfered. All exterior socket joints should face upwards and all interior socket joints should face downwards.
8. Exterior flue pipes should be supported by brackets fitted throughout the height of the flue at intervals of not more than 2 m, preferably immediately below the pipe socket.
9. Where a flue passes through combustible material, it must be covered by a metal or asbestos cement sleeve, with an annular spacing of 25 mm packed with incombustible material (see Fig. 6.18).

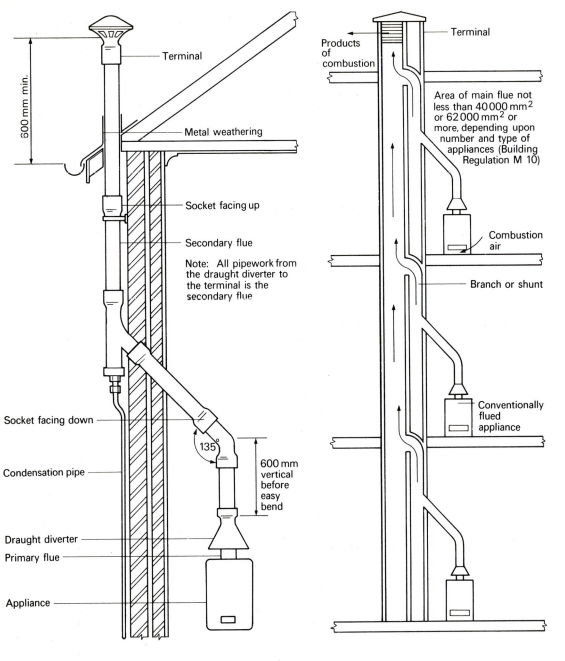

Fig. 6.10 Primary and secondary flue

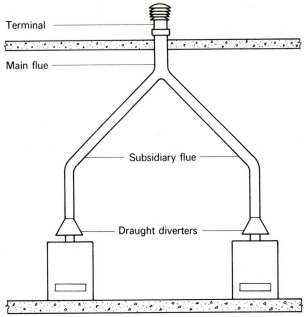

Fig. 6.11 Main and subsidiary flues

Fig. 6.12 Branch and shunt flue

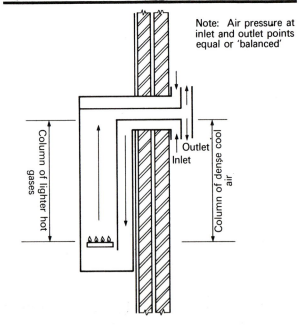

Fig. 6.13 Principle of operation of balanced-flue appliance

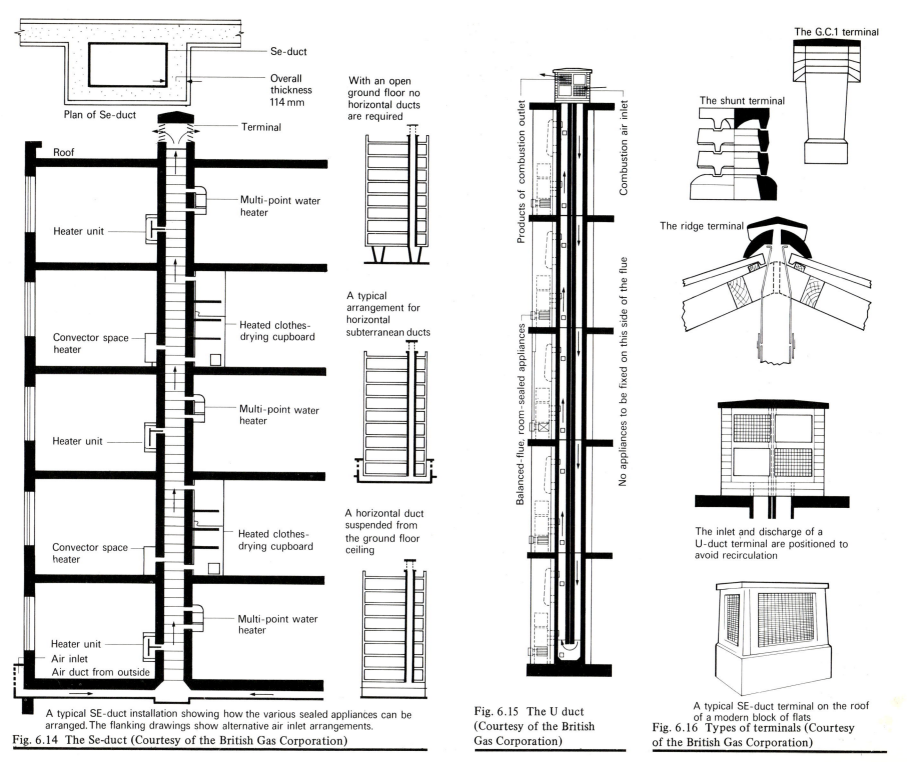

Plan of Se-duct

Se-duct

Overall thickness 114 mm

Roof

Terminal

Multi-point water heater

Heater unit

Convector space heater

Heated clothes-drying cupboard

Multi-point water heater

Heater unit

Convector space heater

Heated clothes-drying cupboard

Multi-point water heater

Heater unit

Air inlet

Air duct from outside

With an open ground floor no horizontal ducts are required

A typical arrangement for horizontal subterranean ducts

A horizontal duct suspended from the ground floor ceiling

A typical SE-duct installation showing how the various sealed appliances can be arranged. The flanking drawings show alternative air inlet arrangements.

Fig. 6.14 The Se-duct (Courtesy of the British Gas Corporation)

Products of combustion outlet

Combustion air inlet

Balanced-flue, room-sealed appliances

No appliances to be fixed on this side of the flue

Fig. 6.15 The U duct (Courtesy of the British Gas Corporation)

The G.C.1 terminal

The shunt terminal

The ridge terminal

The inlet and discharge of a U-duct terminal are positioned to avoid recirculation

A typical SE-duct terminal on the roof of a modern block of flats

Fig. 6.16 Types of terminals (Courtesy of the British Gas Corporation)

77

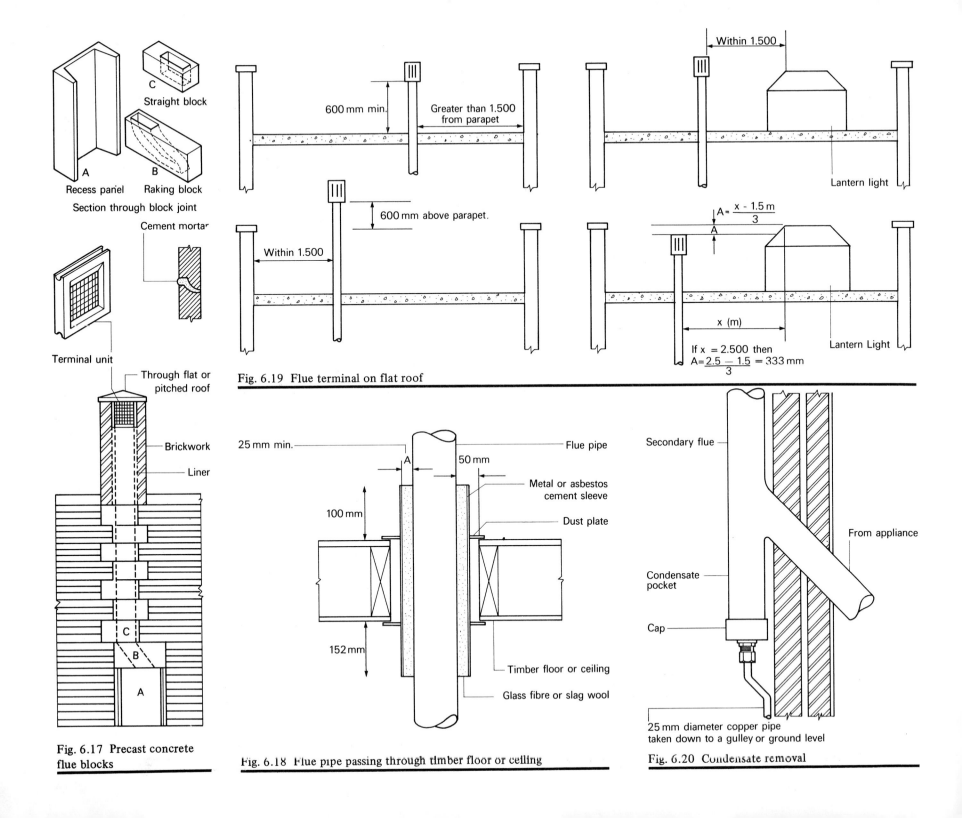

Section through block joint

C — Straight block
A — Recess panel
B — Raking block

Cement mortar

Terminal unit

Fig. 6.17 Precast concrete flue blocks

Through flat or pitched roof
Brickwork
Liner
C
B
A

Fig. 6.19 Flue terminal on flat roof

600 mm min.
Greater than 1.500 from parapet

Within 1.500

Lantern light

600 mm above parapet.

Within 1.500

$A = \dfrac{x - 1.5\,m}{3}$

Lantern Light

x (m)

If x = 2.500 then
$A = \dfrac{2.5 - 1.5}{3} = 333\,mm$

Lantern Light

Fig. 6.18 Flue pipe passing through timber floor or ceiling

25 mm min.
A
50 mm
Flue pipe
Metal or asbestos cement sleeve
Dust plate
100 mm
152 mm
Timber floor or ceiling
Glass fibre or slag wool

Fig. 6.20 Condensate removal

Secondary flue
From appliance
Condensate pocket
Cap
25 mm diameter copper pipe taken down to a gulley or ground level

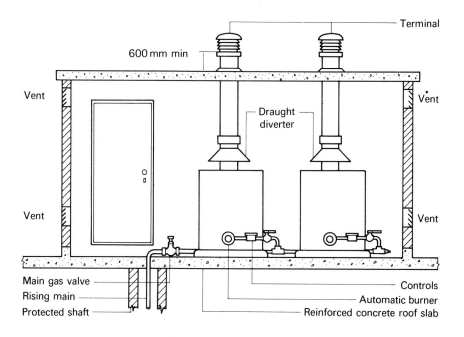

Fig. 6.21 Rooftop boiler room

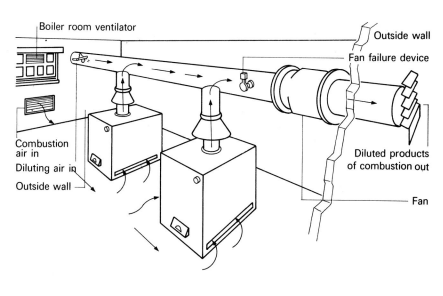

Fig. 6.23 Installation using two outside walls and boilers with draught diverters

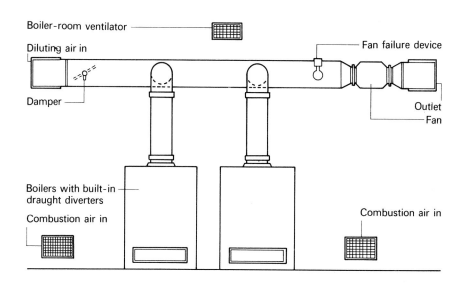

Fig. 6.22, 6.23, and 6.24 Fan-diluted flue (Courtesy of Gas Council)

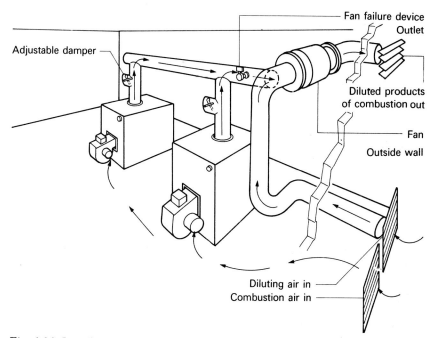

Fig. 6.24 Installation using one outside wall and boilers with automatic gas burners

Terminal position

The terminal of a gas flue should be placed in such a position that the wind can blow freely across it, the best positions being above the ridge of a pitched roof or above the parapet wall of a flat roof. Figure 6.19 shows the requirements under the Building Regulations, 1976 for flues passing through a flat roof.

High-wind-pressure regions must be avoided and therefore terminals must not be placed below the eaves, in a corner or adjacent to another pipe. However, a terminal can be placed above the level of the eaves (as shown in Fig. 6.10) or on the unobstructed portion of a wall.

Condensate and its removal

Some initial condensation in a gas flue will occur immediately after the appliance has been lit and it tends to persist in flues placed on the external walls or in flues attached to appliances of high efficiencies, especially if these are run for long periods at much less than their rated maximum output. Well-insulated internal flues are preferable to external flues which increase the degree of condensation. The flues should be built so that condensate can flow freely to a point where it can be released, preferably into a 25 mm bore lead or copper condensate pipe (see Fig. 6.20).

Large boilers

It is desirable to house large gas-fired boilers in separate boiler rooms, which may be part of the main structure or completely separate. A boiler room sited on the roof of the building requires a very short flue (see Fig. 6.21). Alternatively, the boiler may be installed at ground level and a fan-diluting flue used (see Figs. 6.22, 6.23 and 6.24). With this type of flue, a fan draws in fresh air which is mixed with the products of combustion from the appliance and discharged to the outside air.

It is essential to provide permanent air inlets to the boiler room to ensure a sufficient supply of air for the efficient operation of the boilers and these inlets should be:

1. At least twice as great in free area as the area of the primary flue pipe.
2. Located at least 300 mm above ground level and fitted with a grill.
3. Constructed of durable material.

Note: The provision of combustion air to gas appliances, is covered by the Building Regulations, 1976 in M8 and M12.

Shared flues

Gas appliances in multi-storey buildings may be connected to a shared flue in the form of a SE-duct, U duct or shunt duct. The use of a shared flue in a multi-storey building, saves considerably in space and installation costs over the use of individual flues. Both the SE-duct and the U duct require room-sealed, balanced-flue appliances, and the shunt duct requires a conventional flue appliance which has the advantage of ventilating the room in which the appliance is installed. For safety reasons, all appliances connected to shared flues must be provided with flame-failure devices.

The degree of dilution of the products of combustion in a shared flue is sufficient to ensure the satisfactory operation of all gas appliances connected to the system.

Calculation of gas consumption

The gas requirement from appliance may be found from the following formula:

$$\frac{power}{calorific\ value} = m^3/s$$

The calorific or heating value of natural gas is approximately 37 000 kJ/m³ and town gas approximately 19 000 kJ/m³.

Example 6.1. *Calculate the gas consumed in cubic metres per hour by a 20 kW boiler when natural gas is used.*

$$Consumption = \frac{20 \times 3600}{37\ 000} = 1.945\ m^3/h$$

Note: 1 kW   = 3.6 MJ/h
     1 Therm = 105.5 MJ

Electric space heating

Electric space heating can be broadly classified into direct and indirect systems. Direct systems use appliances such as fires and convector heaters, using electricity at the standard rate; they have the advantage of flexibility, immediate response to heating, and utilise 100 per cent of their input of power.

Indirect or storage systems are operated on 'off-peak' supplies and are therefore cheaper to operate than direct systems but require some time, due to thermal lag, before they can be used for heating. Indirect systems include block-type storage heaters and underfloor heating. The direct and indirect systems may be used independently, or may be complementary to one another to meet any particular heating requirement.

Advantages of electric heating

1. Absence of products of combustion or fumes near the building.
2. Clean and silent.
3. Economical in builders' work, as no fuel storage, boiler room and chimney are necessary.
4. Ease of maintenance and automatic control.
5. Portable appliances may be used.
6. No risk of freezing or leaks.
7. Pumping not required and heating may be achieved at any point regardless of height.

Direct systems

A wide range of direct-heating electrical appliances are available such as fires, convectors, tubular heaters, oil-filled radiators, fan heaters, and high-temperature panels. The appliances can be fixed or portable and some are available in both forms. Direct electric heating are normally more expensive to run than indirect storage heating, but is easier to control. Electric fires, tubular heaters, convectors and overhead radiant heaters, are commonly used in offices, factories, shops and residential buildings. They can also be used for 'topping-up' where storage systems provide background heating. Figures 6.25–6.30 show various types of direct-heating appliances.

Ceiling heating

Its low thermal capacity makes an electrically heated ceiling very flexible and convenient, as it responds quickly to thermostatic and time control. Because the ceiling provides radiant heating, a room is comfortably warm when the air temperature is only 17 °C to 18 °C. When the heating is by convection, an air temperature of 20 °C to 21 °C is required to provide the same level of thermal comfort. The lower air temperature of radiant heating systems means that heat losses from air escaping through open doors and windows is reduced. The occupants of the room also feel fresher.

Ceiling heating is suitable for most well-insulated buildings, including shops, hotels, flats, houses, churches and hospitals. The system consists of a special flexible sheet material, made by coating glasscloth with a conducting silicone elastomer. Each ceiling heating element is fitted with copper strip electrodes along each side and enclosed in an electrically insulating sheath. The heating elements are installed immediately above the ceiling face, usually by stapling them to joists or battens. A layer of mineral wool or glassfibre is laid above the elements to provide insulation (see Figs. 6.31 and 6.32). Connections to the mains electricity supply — and between individual elements — is made from one end of the elements only so that only one supply point is needed in each room. A thermostat is incorporated in the circuit. The power loading depends upon the height of the ceiling and the dimensions of the room, but it is usually between 200 W and 250 W/m^2. The ceiling temperature is between 32 °C and 38 °C, and in order to avoid too high a temperature at head level, the ceiling height should not be less than 2.3 m.

Indirect storage systems

In order to encourage the use of electricity when power stations are operating on low loads, the Electricity Boards offer low-rate tariffs for power consumed during these off-peak times, usually between 19.00 h and 07.00 h. The power consumed is recorded by a white meter and a time switch controls the time when the current is switched ON and OFF. The heat given out by the electrical elements is stored during the time the current is switched ON and this heat is released when the current is switched OFF. The main types of indirect storage appliances are as follows:

1. *Block storage heaters:* (*a*) The heat is stored in refectory blocks contained in an insulated metal casing and given out by radiation and natural convection (see Fig. 6.33).

2. *Block storage heaters:* (*b*) The heat is stored in refectory blocks contained in an insulated metal casing and given out again by radiation and forced air flow produced by a fan (see Fig. 6.34).

3. *Warm-air system:* A block storage heater with a fan is used and the heat forced by the fan through ductwork to the various rooms (see Fig. 6.35).

4. *Hot-water thermal storage system:* 'Off-peak' electrical energy is converted into heat energy and used to heat water, which is transferred to well-insulated horizontal or vertical thermal storage cylinders (see Fig. 6.36).

The heat generator can be an electrode boiler shown in Fig. 6.37, or an electric immersion heater inside the cylinder. Operation is as follows:

(*a*) The electrode boiler or immersion heater heats the water in the cylinder.

(*b*) This heat takes place during the 'off-peak' period, and in order to reduce the size of the storage cylinder the water temperature is raised as high as possible without generating steam. The water temperature may be as high as 185 °C, depending upon the pressure of water acting upon the cylinder. The water pressure may be increased by the use of a gas cushion inside a cylinder connected to the return pipe.

(*c*) On completion of the heating period, the current to the heater is cut off automatically by a switch and there is sufficient heat energy in the storage cylinder to maintain the building's space heating and hot-water requirements until the next period.

5. *Floor-warming system:* Electric heating cables are either directly embedded in the floor screed, or passed through metal conduit which is also embedded in the screed. Most floor-warming systems, employ the direct embedded-type element (see Fig. 6.38). The concrete floor provides the necessary heat storage, but it is essential that insulation is incorporated in the floor slab to prevent loss of heat. Installations are designed to provide an average floor surface temperature of 24 °C and the loading per square metre is between 100 and 150 W. The heating cables should cover the whole floor at spacings not more than 100 m or less than the screed depth, so as to avoid temperature variations on the floor surface.

Calculations

The heat energy or the specific heat capacity (s.h.c.) of water is equal to 4.2 kJ/kg °C. In other words, it requires 4.2 kJ of heat energy to raise the temperature of 1 kg of water through 1 °C. Also

$$kW = \frac{kJ}{seconds} \quad or \quad \frac{s.h.c.}{seconds}$$

Example 6.2 *Calculate the power in kW required to raise the temperature of 136 kg of water from 10 °C to 60 °C in 2 h, when the heat losses are 20 per cent.*

$$kW = \frac{s.h.c. \times temperature\ rise\ °C \times kg \times 100}{heating\ time\ in\ seconds \times efficiency}$$

$$kW = \frac{4.2 \times (10 - 60) \times 136 \times 100}{2 \times 3600 \times 80}$$

$$kW = 4.95$$

A 5 kW electric immersion heater would be required.

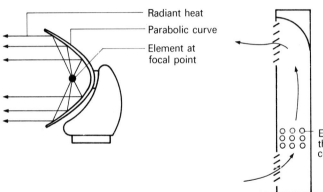

Fig. 6.25 Electric fire with parabolic reflector

Fig. 6.26 Electric natural convector heater

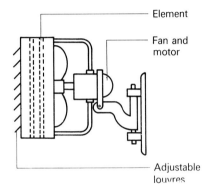

Fig. 6.27 Electric unit heater

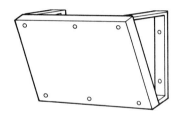

Fig. 6.28 High-temperature electric panel heater

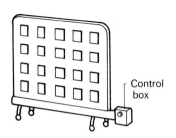

Fig. 6.29 Oil-filled portable radiator

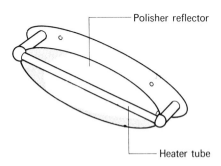

Fig. 6.30 Wall-mounted infrared heater

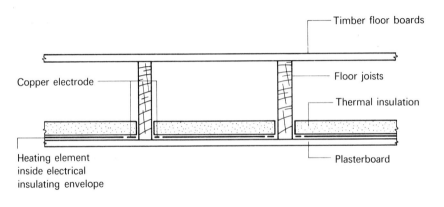

Fig. 6.31 Ceiling heating in timber floor

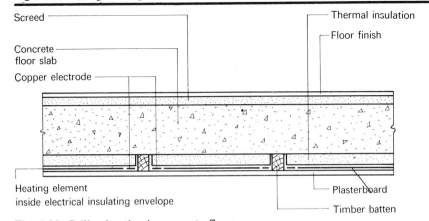

Fig. 6.32 Ceiling heating in concrete floor

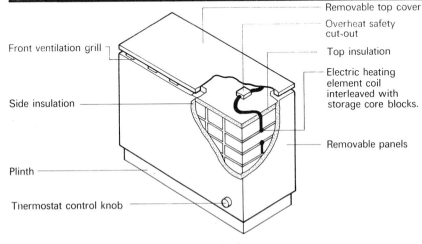

Fig. 6.33 Block storage heater (without fan)

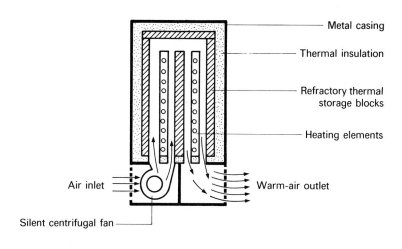

Fig. 6.34 Block storage heater (with fan)

Metal casing

Thermal insulation

Refractory thermal storage blocks

Heating elements

Air inlet

Warm-air outlet

Silent centrifugal fan

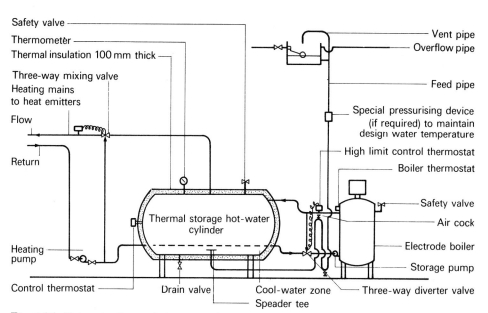

Safety valve

Thermometer

Thermal insulation 100 mm thick

Three-way mixing valve
Heating mains to heat emitters

Flow

Return

Heating pump

Control thermostat

Vent pipe

Overflow pipe

Feed pipe

Special pressurising device (if required) to maintain design water temperature

High limit control thermostat

Boiler thermostat

Safety valve

Air cock

Electrode boiler

Storage pump

Three-way diverter valve

Thermal storage hot-water cylinder

Drain valve

Cool-water zone

Speader tee

Fig. 6.36 Hot-water thermal storage system

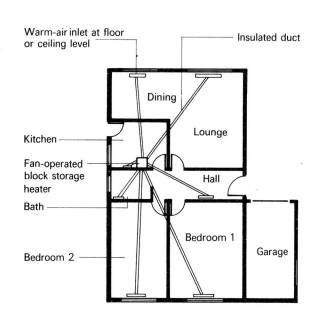

Warm-air inlet at floor or ceiling level

Insulated duct

Kitchen

Fan-operated block storage heater

Bath

Bedroom 2

Dining

Lounge

Hall

Bedroom 1

Garage

Fig. 6.35 Electric warm-air heating system (with radial ducts)

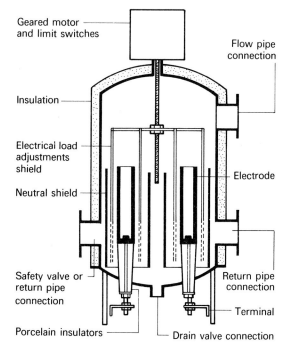

Geared motor and limit switches

Insulation

Electrical load adjustments shield

Neutral shield

Safety valve or return pipe connection

Porcelain insulators

Flow pipe connection

Electrode

Return pipe connection

Terminal

Drain valve connection

Fig. 6.37 Electrode boiler

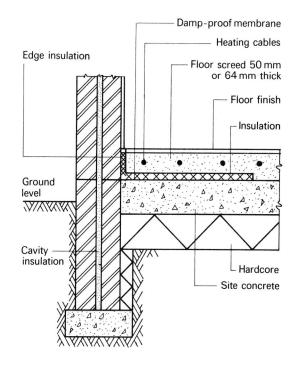

Edge insulation

Ground level

Cavity insulation

Damp-proof membrane

Heating cables

Floor screed 50 mm or 64 mm thick

Floor finish

Insulation

Hardcore

Site concrete

Fig. 6.38 Electric underfloor heating

Example 6.3 *Calculate the time taken in hours, and the cost in pence, of heating 227 kg of water from 10 °C to 60 °C by means of a 5 kW immersion heater when the heat losses are 20 per cent and 1 kWh costs 2p.*

$$kW = \frac{\text{s.h.c.} \times \text{kg} \times \text{temperature rise} \,^{\circ}C \times 100}{\text{heating time in seconds} \times \text{efficiency}}$$

By transposition:

$$\text{Time in seconds} = \frac{\text{s.h.c.} \times \text{kg} \times \,^{\circ}C \times 100}{kW \times \text{efficiency}}$$

$$\text{Time in seconds} = \frac{4.2 \times 227 \times 50 \times 100}{5 \times 80}$$

Time in seconds = 11 917.5

$$\text{Time in hours} = \frac{11\,917.5}{3600}$$

Time in hours = 3.31 h

Cost = kWh × 2

     = 5 × 3.31 × 2

     = 33.1p

Example 6.4 *An immersion heater has a resistance of 20 Ω and has a current passing through of 12 A. If the heat losses are 20 per cent, calculate the mass of water in kilograms that can be heated from 5 °C to 60 °C in 1 h.*

Volts = current × resistance

     $V = AR$

     $V = 12 \times 20$

     $V = 240$

Watts = volts × amps

     $W = VA$

     $W = 240 \times 12$

     $W = 2880$

At 80 per cent efficiency:

$$W = 2880 \times \frac{80}{100}$$

     $W = 2304$

∴ $kW = 2.3$ (approx.)

     kJ = kW × seconds

     kJ = 2.3 × 3600

     kJ = 8280

Heat in water in kilojoules = 8280. Therefore

8280 = kg × 4.2 × temperature rise °C

8280 = kg × 4.2 × 55

$$kg = \frac{8280}{4.2 \times 55}$$

     = 35.84 kg, say 36 kg

Chapter 7

Heat gains and solar heating

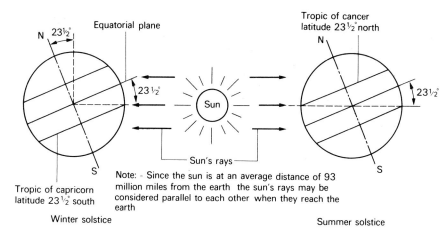

Fig. 7.1   Winter and summer solstice

Solar heat gains (principles)

In order to understand calculations involved in determining heat gains in a building due to solar radiation, a knowledge of the relative movements of the earth and sun are essential.

The earth rotates about its axis once every 24 hours and revolves about the sun once every year. The axis of the earth is inclined at an angle of $23.5°$ to the ecliptic. At the winter solstice, i.e. 21 December, the North Pole is inclined away from the sun and at the summer solstice, i.e. 21 June, the North Pole is inclined towards the sun. These changes are due to movement of the earth in its orbit around the sun and the two positions are shown in Fig. 7.1 .

By inspection of Fig. 7.1 it will be noticed that the sun will be seen directly overhead at different earth latitudes. On 21 December, the sun will be seen directly overhead on the Tropic of Capricorn and on 21 June, it will be directly overhead on the Tropic of Cancer.

Countries to the south of the Tropic of Capricorn and to the north of the Tropic of Cancer will, however, never have the sun appearing directly overhead. At the spring and autumn equinoxes, i.e. 21 March and 21 September, the earth's poles are equidistant from the sun, so that everywhere on the earth's surface has 12 hours of daylight and 12 hours of darkness. The sun also appears directly overhead at the equator on these days (see Fig. 7.2).

Monthly declination

Since four fixed positions of the earth relative to the sun are known, it is

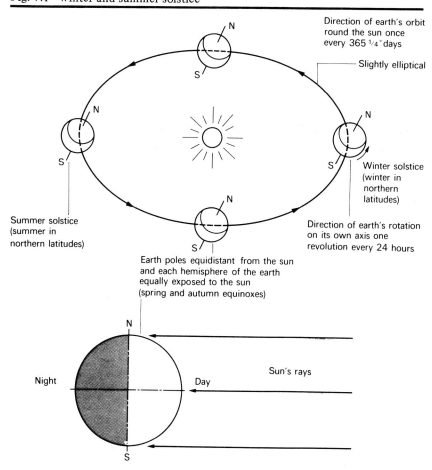

Fig. 7.2   Relative positions of earth and sun–spring and autumn equinoxes

85

possible to calculate the latitudes on the earth where the sun will appear directly overhead for the remaining months of the year (see Fig. 7.3).

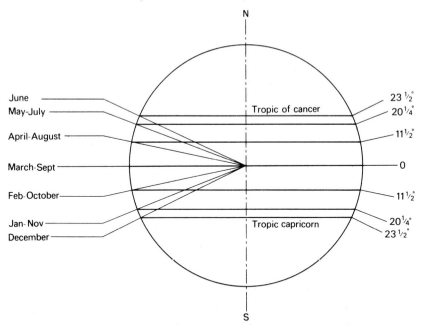

Fig. 7.3 Approximate angles of declination on the 21st of each calendar month

If required, it would be possible to calculate the relative position of the sun for every day of the year. For the purposes of solar heat gain, however, the designer is normally interested only in information for the period of maximum heat gain, which will be when the sun is at its highest position in the sky.

For latitudes lying between $23.5°$ north and $23.5°$ south, this would be when the sun is directly overhead. For latitudes outside these limits, however, the altitude of the sun must be calculated.

Example 7.1. *Calculate the maximum and minimum altitudes of the sun in Birmingham, which lies on latitude $52.5°$ north.*

Altitude is the angle the sun's rays make with the tangent to the earth at the altitude being considered. The greatest altitude will occur at noon on 21 June, and the least at noon on 21 December. By knowing that the declinations of the sun are $23.5°$ N and $23.5°$ S respectively, and assuming that the sun's rays are parallel, the maximum altitude can be illustrated by a diagram (see Fig. 7.4).

The following expressions for maximum and minimum altitudes can be obtained.

maximum altitude $(a) = 90 - (L - d)$

minimum altitude $(a) = 90 - [L - (-d)]$

where L = latitude being considered

d = maximum declination of the sun

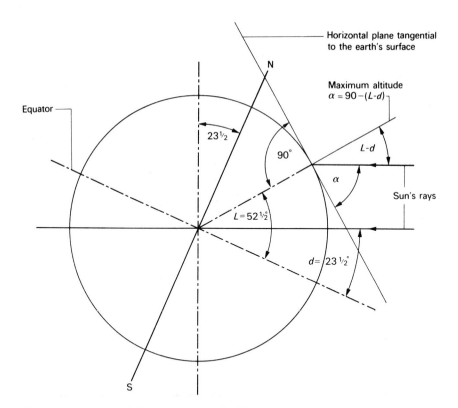

Fig. 7.4 Altitude of the sun in example 7.1

∴ the answers will be

maximum altitude $(a) = 90 - (52.5 - 23.5)$

$= 61°$

minimum altitude $(a) = 90 - [52.5 - (-23.5)]$

$= 14°$

Sun's position

Since the earth is rotating about its axis once every 24 hours, these altitudes will only be observed at noon; noon in this context being the time of day when the sun lies due south of the observer.

Solar azimuth angle (see Fig. 7.5)

For northern latitudes the solar azimuth angle is the angle in the horizontal plane that the horizontal component of the sun's rays make with the true north when measured from the true north in a clockwise direction. For southern latitudes, the azimuth angle is measured from the south in a clockwise direction.

Figure 7.5 shows the solar azimuth angle for northern latitudes.

It is possible, by the use of three-dimensional trigonometry, to relate the declination latitude, altitude and azimuth angles, and to calculate any of these

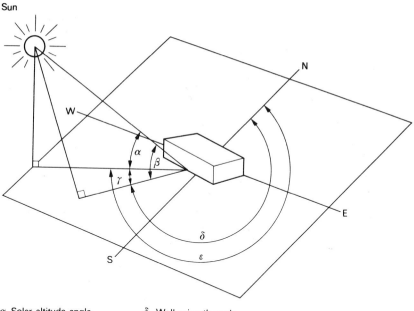

α Solar altitude angle
β Angle of incidence
γ Wall-solar azimuth angle

δ Wall azimuth angle
ε Solar azimuth angle
   (for northern latitudes)

Fig. 7.5 Definition of angles

for any point on the earth's surface at any time. The result, obtained from such trigonometrical consideration is as follows:

$$\sin a = (\sin d \sin L) + (\cos d \cos L \cos b)$$

where a = the altitude of the sun

d = angle of declination

L = latitude

b = hour angle

Example 7.2 . *Calculate the altitude of the sun for London at 10.00 hours sun time on 21 June, given that the latitude for London is 51.5°.*

Note: Sun time is the time in hours, before or after noon; noon in this context being the time when the sun is highest in the sky.

The angle of declination, d, is 23.5° and L is 51.5° N. Since the earth revolves once every 24 hours, it will move through an angle of 360/24 = 15° in 1 hour. This angle is measured from either side of noon sun time, and therefore 10.00 hours represents an hour angle of 2 × 15° = 30°, hence:

$$\sin a = (\sin 23.5° \sin 51.5°) +$$
$$(\cos 23.5° \cos 51.5° \cos 30°$$

$$= (0.3987 \times 0.7826) +$$
$$(0.9171 \times 0.6225 \times 0.8660)$$
$$= 0.8063$$
$$\therefore a = 53.7° \text{ (approx.)}$$

If the solar azimuth angle, ϵ, measured east or west of south, is required, the following formula may be applied:

$$\tan \epsilon = \frac{\sin b}{(\sin L \cos b) - (\cos L \tan d)}$$

$$= \frac{\sin 30°}{(\sin 51.5° \cos 30°) - (\cos 51.5° \tan 23.5°)}$$

$$= \frac{0.5}{(0.7826 \times 0.8660) - (0.6225 \times 0.4348)}$$

$$= 1.25$$

$$\therefore \epsilon = 51° \text{ E or W of S (approx.)}$$

depending upon the orientation of the building.

$$\text{also } \epsilon = 180° - 51° = 129°$$
$$\text{or } \epsilon = 180° + 51° = 231°$$

Similar calculations may be carried out to obtain values for the altitude and solar azimuth angles for various latitudes and sun times. The Chartered Institution of Building Services (CIBS) *Guide Book A,* Section 6, provides tables of solar altitude and azimuth angles for various sun times of each month.

Sun angles

The position of the sun in relation to the orientation of any surface of a building may be specified by the angles shown in Fig. 7.5 , and these include the altitude and solar azimuth angles previously described. The angle of incidence on the building face (β) is that angle between the sun's direction and the normal to the wall and is given by the following formula:

$$\cos \beta = \cos a \cos \gamma$$

where $\cos \beta$ = the angle of incidence

$\cos a$ = the altitude of the sun

$\cos \gamma$ = the wall-solar azimuth

Intensity of solar radiation

The average intensity of solar radiation normal to the sun's rays at the outer edge of the earth's atmosphere is 1362 W/m² and is subject to seasonal variation of + 3.5 per cent in January and − 3.5 per cent in July. In passing through the atmosphere, part of the heat is absorbed, part scattered into space, and part scattered back to earth by the atmosphere. The portion which comes directly through the atmosphere is termed direct radiation and the portion scattered back to earth from the atmosphere is termed sky diffuse radiation. For cloudless skies, the intensities of direct and diffuse radiation will depend upon the thickness of the layer of atmosphere traversed by the sun's rays and on the solar altitude and the height above sea-level. They also depend upon the proportions of water vapour, dust and ozone in the atmosphere, which scatter and absorb radiation.

Vertical surfaces

The intensity of direct radiation on a vertical surface facing due south at noon sun time is given by the following expression:

$$I_s = I \cos a$$

where
I_s = the direct incident radiation on the surface (W/m²)
I = the intensity of direct radiation (W/m²)

Figure 7.6 shows the intensity of direct radiation on a vertical surface facing south at noon sun time.

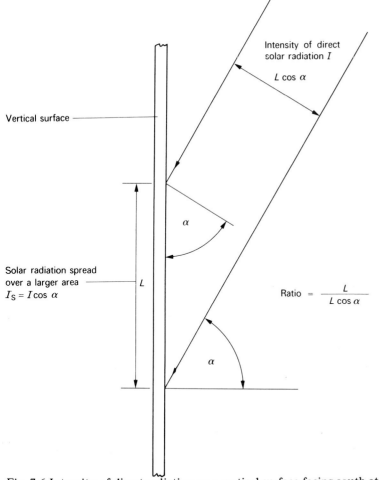

Intensity of direct solar radiation I

$L \cos \alpha$

Vertical surface

Solar radiation spread over a larger area
$I_S = I \cos \alpha$

L

α

α

Ratio $= \dfrac{L}{L \cos \alpha}$

Fig. 7.6 Intensity of direct radiation on a vertical surface facing south at noon sun time

Hence, the heating effect at noon on a vertical surface facing south will be given by:

$$\text{heating effect} = \text{area} \times I \times \cos a \qquad (\text{W})$$

Example 7.3. *Calculate the total heating effect of the sun on a vertical window which faces south given the following data:*

time of day	*noon*
window size	*3 m × 2 m*
latitude	*56¼° N*
day and month	*21 July*
intensity of direct radiation normal to sun	*620 W/m²*

By reference to Fig. 7.3, on 21 July, the declination of the sun is 20¼°.

altitude of the sun (a) = 90 − (L − d)

= 90 − (56¼ − 20¼)

= 54°

heating effect = area × I × cos a

= 3 × 2 × 620 × cos 54

= 6 × 620 × 0.5878

= 2186.561 W

Note: The intensity of direct radiation normal to the sun may be obtained from the CIBS *Guide Book A* or from the Meteorological Office.

Example 7.4. *Calculate the total heating effect of the sun in London on a vertical window which faces south given the following data:*

time of day	*noon*
window size	*2 m × 1.5 m*
latitude	*51.7° N*
day and month	*21 June*
intensity of direct radiation normal to sun	*850 W/m²*

Declination of the sun on 21 June is 23½°.

altitude of the sun (a) = 90 − (L − d)

= 90 − (51.7 − 23.5)

= 61.8°

heating effect = area × I × cos a

= 2 × 1.5 × 850 × 0.4726

= 1205.13 W

Horizontal surfaces

Similar calculations may be carried out to determine the heating effect on horizontal surfaces, and the following formula may be used:

$$\text{heat load} = \text{area} \times I \times \sin a$$

Example 7.5. *Calculate the total heating effect of the sun in London on a horizontal skylight using the same values given in example 7.4.*

$$\text{heat load} = \text{area} \times I \times \sin a$$
$$= 2 \times 1.5 \times 850 \times 0.8813$$
$$= 2247.315 \text{ W}$$

Note: The total heating effect on the horizontal surface has been increased by 1042.185 W. This is because solar radiation is spread over a smaller area on the horizontal surface, which increases its intensity (for the same altitude of the sun).

Sloping surfaces

Sloping surfaces such as pitched roofs provide additional problems of solar heat-gain calculations due to the introduction of another angle, i.e. roof slope. Two formulae have been derived for the calculation of the incident radiation on the surface, depending upon whether the roof is facing into or away from the sun.

1. Roof facing into the sun:

$$I_s = (I \sin a \cos \delta) + (I \cos a \cos \gamma \sin \delta)$$

2. Roof facing away from the sun:

$$I_s = (I \sin a \cos \delta) - (I \cos a \cos \gamma \sin \delta)$$

where I_s = incident radiation (W/m^2)

δ = angle of roof slope

a = altitude of the sun

γ = wall-solar azimuth angle

Example 7.6 illustrates the application of the formulae.

Example 7.6. *Find the intensity of solar radiation on a sloping roof of a building situated at latitude 50° N. The orientation of the ridge is 19° N of W to 19° S of E and the roof slopes at 20°. The time may be taken at 14.00 hours on 21 July.*

By reference to Fig. 7.3, the declination of the sun on 21 July is 20.25°. The hour angle for noon sun time plus two hours will be $2 \times 15° = 30°$.

$$\therefore \sin a = (\sin 20.25° \times \sin 50°) + (\cos 20.25° \times \cos 50° \times \cos 30°)$$
$$= (0.3461 \times 0.7660) + (0.9382 \times 0.6428 \times 0.8660)$$
$$= 0.7874$$

solar altitude (a) = 52° (approx.)

The solar azimuth angle ϵ may now be calculated as follows:

$$\tan \epsilon = \frac{\sin 30°}{(\sin 50° \times \cos 30°) - (\cos 50° \times \tan 20.25°)}$$
$$= \frac{0.5}{(0.7660 \times 0.8660) - (0.6428 \times 0.3689)}$$
$$= 1.1731$$

$\therefore$ solar azimuth angle ϵ = 49° W of S (approx.)

or $\epsilon = 49 + 180 = 229°$ SW of N

The wall-solar azimuth angle may now be found as shown in Fig. 7.7.

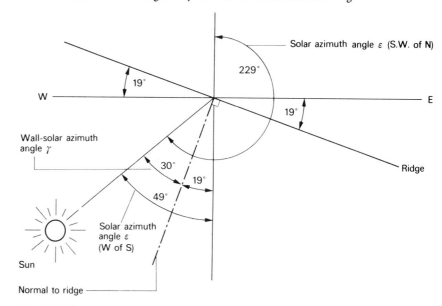

Fig. 7.7 Method of finding the wall-solar azimuth angle

Since the roof is facing into the sun and inclined at 20° to the sun, the intensity of direct solar radiation normal to the sun may be taken as being about 800 W/m^2. The intensity of direct radiation on the roof on the same side of the ridge may now be found as follows:

$$I_s = (I \sin \alpha \cos \delta) + (I \cos a \cos \gamma \sin \delta)$$
$$= (800 \sin 52° \cos 20°) + (800 \cos 52° \cos 30° \sin 20°)$$
$$= (800 \times 0.7880 \times 0.9397) + (800 \times 0.6157 \times 0.8660 \times 0.3420)$$
$$= 592.386 + 145.882$$
$$= 738.3 \text{ W/m}^2 \text{ (approx.)}$$

The incident radiation on the roof on the opposite side of the ridge to the sun will be as follows:

$$I_s = (I \sin a \cos \delta) - (I \cos a \cos \gamma \sin \delta)$$
$$= (800 \sin 52° \cos 20°) - (800 \cos 52° \cos 30° \sin 20°)$$

$$= 592.386 - 145.882$$

$$= 446.5 \ \text{W/m}^2 \ (\text{approx.})$$

Heat transmission through glass

The values of incident radiation so far obtained provide an estimate of the quantity of heat incident upon the outside surface of a structure. Not all this heat, however, is transmitted to the inside of the building; some is reflected or absorbed by the glass. The thermal capacity of the building fabric also has an important effect on solar heat gains. The radiation is usually incident on the floor where part is absorbed and part is diffusely reflected, to be subsequently absorbed by the walls and the ceiling.

Example 7.7. *If the roof in example 7.6 contains a skylight of single glazing measuring 1.5 m by 1 m, calculate the proportion of heat transmitted to the building using the following factors:*

$$\text{glass transmission factor} = 0.46$$

$$\text{glass absorption factor} = 0.40$$

percentage of absorbed radiation transmitted to the room $= 68$ *per cent*

$$\text{Roof} \ I_s = 738.3 \ \text{W/m}^2$$

$$\text{heating effect on glass} = 738.3 \times 1.5 \ \text{m}^2$$

$$= 1107.45 \ \text{W}$$

$$\text{portion transmitted to the room} = 1107.45 \times 0.46$$

$$= 509.43 \ \text{W}$$

$$\text{portion absorbed by the glass} = 1107.45 \times 0.4$$

$$= 442.98 \ \text{W}$$

$$\text{absorbed energy transmitted to the room} = 442.98 \times 0.68$$

$$= 301.23 \ \text{W}$$

$$\text{total heat transmitted to the room} = 509.43 + 301.23$$

$$= 810.66 \ \text{W}$$

Effect of other orientations

So far, consideration of solar heat gains have been given for surfaces facing due south, i.e., buildings having an east–west line.

In order to calculate heat gains for buildings having other orientations, the wall-solar azimuth angle γ must be included, which has been described and calculated previously.

The value of the direct solar radiation upon vertical surfaces of any orientation for any time of day is found from the following formula:

$$I_s = \cos a \cos \gamma$$

where I_s = the direct incident radiation on the surface (W/m^2)

I = the intensity of direct radiation (W/m^2)

a = solar altitude angle

γ = wall-solar azimuth angle

Hence the heating effect on the surface will be:

$$\text{heating effect} = \text{area} \times I \cos a \times \cos \gamma \qquad (\text{W})$$

Example 7.8. *Calculate the total heating effect of the sun on a vertical window which faces southeast given the following data:*

time of day	*noon*
altitude of sun	*58°*
latitude	*54° N*
wall-solar azimuth angle	*61° E of S*
intensity of direct radiation normal to sun	*800 W/m²*
window size	*2 m × 1.5 m*

$$\text{heating effect} = \text{area} \times I \cos 58° \times \cos 61°$$

$$= 2 \times 1.5 \times 800 \times 0.5299 \times 0.4848$$

$$= 616.5 \ \text{W} \ (\text{approx.})$$

This heating effect will be reduced before it enters the room, as previously described in example 7.7.

Effect of shading

Most window frames are recessed into the wall and this causes shading of the glass and a reduction in solar heat gain into the room. A large recess is sometimes used in order to achieve this reduction in heat gain and glare.

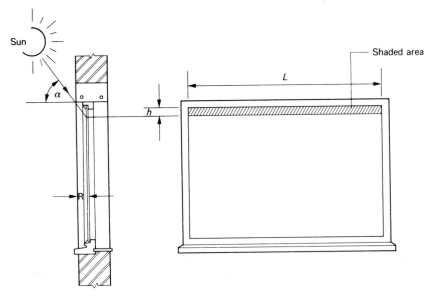

Fig. 7.8 Window facing due south – depth of shade at noon sun time

For a window facing directly into the sun (which will give the highest heat gain), the shaded area can be calculated from the altitude angle of the sun (see Fig. 7.8) and the shaded area is equal to:

$$\text{shaded area} = L \times h$$

$$= L \times R \tan a$$

where L = length of window in (m)

a = altitude angle of the sun

R = depth of recess in (m)

Example 7.9.   *A window frame facing directly into the sun measures 1.5 m long × 1 m high. If the altitude angle of the sun is found to be 50° and the solar radiation intensity 860 W/m², calculate the heating effect on the unshaded area of the glass when the depth of recess is 300 mm.*

shaded area of glass = $L \times R \tan a$

$= 1.5 \times \tan 50° \times 0.300$

$= 1.5 \times 1.1918 \times 0.300$

$= 0.5363 \text{ m}^2$

unshaded area of glass = $1.5 - 0.5363$

$= 0.9637 \text{ m}^2$

heating effect on
unshaded area of glass = 0.9637×860

$= 828.782 \text{ W}$

If the window is not facing directly into the sun, calculation of the shaded area involves another angle and therefore the heating effect is more difficult to determine. In Fig. 7.9 the angle γ causes the sun's rays to travel a horizontal distance y so that the depth of shading is given by:

$$h = y \times \tan a$$

$$y = \frac{R}{\cos \gamma}$$

$$\therefore h = \frac{R \tan a}{\cos \gamma}$$

The amount of side shading is given by:

$$x = R \times \tan \gamma$$

Example 7.10.   *A window, not facing directly into the sun, measures 2 m long × 1.5 m high. If the altitude angle of the sun and the angle γ are found to be 38° and 36° respectively, calculate the heating effect on the unshaded area of glass when the solar radiation intensity is 750 W/m² and the depth of recess 150 mm.*

$$h = \frac{0.15 \, (\tan 38°)}{\cos 36°}$$

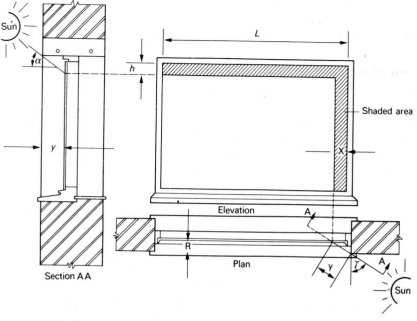

Fig. 7.9 Window in shade for any orientation

$$= \frac{0.15 \times 0.7813}{0.8090}$$

$$= 0.145 \text{ m (approx.)}$$

and $x = 0.15 \times \tan 36°$

$$= 0.15 \times 0.7265$$

$$= 0.11 \text{ m (approx.)}$$

Therefore the total area of the window in the shade would be equal to:

$$(2 \times 0.145) + [(1.5 - 0.145) \times 0.11]$$

$$= (0.29 + 0.149)$$

$$= 0.439 \text{ m}^2$$

$\therefore$ unshaded area = $(2 \times 1.5) - 0.439$

$$= 2.561 \text{ m}^2$$

heating effect on
unshaded area of glass = 2.561×750

$$= 1920.75 \text{ W}$$

Alternatively, the area of unshaded glass may be obtained from:

$$= (1.5 - h) (2 - x)$$

$$= (1.5 - 0.145) \times (2 - 0.11)$$

$$= 1.355 \times 1.89$$

$$= 2.57 \text{ m}^2$$

91

which is approximately equal to the above answer, i.e. 2.561 m²

Note: The heating effect on the unshaded area of glass due to direct solar radiation does not include the diffused or ground-reflected radiation which has also to be considered regardless of whether the glass is in the shade.

In heavy industrial areas, where buildings are close together, the permanent smoke haze may give rise to a greater heat gain by diffused radiation than that obtained from direct solar radiation.

Heat gains through building structure

Heat gains through glass are instantaneous, whereas heat gains through the fabric of a building are delayed due to the thermal time-lag caused by the building material. Different materials will produce different thermal time-lags and heavy-weight constructed buildings have a high thermal capacity and will take longer to cool down or heat up than lightweight constructed buildings. If similar conditions exist, however, a heavyweight constructed building will require more heat input than a lightweight constructed building to bring it back to thermal comfort level.

The calculation of the maximum solar heat gain into a room therefore must include the instantaneous heat gain through the window, plus the fabric heat gain at some time previously, depending upon the thermal time-lag. To allow for this reduction in peak value, it is necessary to determine the gain to the external wall and then apply what is termed a 'decrement factor' (see Table 7.1).

Table 7.1 Adjustment factors to solar heat gains through building fabric

Construction	Adjustments time-lag (hours)	Decrement factor
Light frame (internally lined)	$\frac{1}{2}$	1.0
105 mm brickwork (internally lined)	4	0.7
220 mm brickwork (internally lined)	$8\frac{1}{2}$	0.3
150 mm concrete (internally lined)	$5\frac{1}{2}$	0.5
200 mm concrete (internally lined)	$6\frac{1}{2}$	0.4

Sol-air temperature

The calculation of the gain to the outside surface of the fabric is complicated by the fact that both the incidence solar radiation and the outside air temperature must be considered. This calculation, however, may be simplified by the concept of what is termed the 'sol-air' temperature.

The 'sol-air' temperature is defined as the theoretical outside temperature which would result in the same rate of heat transfer through the structure as exists with the actual solar radiation and the outside air temperature. The sol-air temperature at a certain time may be calculated from the following expression:

$$t_e = t_{ao} + \frac{aI}{b_{so}}$$

where
t_e = sol-air temperature

t_{ao} = actual outside air temperature

a = absorption coefficient applied to the outside surface of the building material

I = intensity of direct plus diffused solar radiation on the outside surface

b_{so} = heat transfer coefficient for the external surface

In practice, it is unnecessary to calculate the value of the sol-air temperature, as these are listed in the CIBS *Guide Book A*, Section 6.

Casual heat gains

Internal heat gains

In addition to heat gains from external sources, there will also be sensible or latent heat gains (or both) within the space itself. These gains are as follows: due to: occupants, lighting, electrical machinery, gas appliances and cooking.

Occupants: The heat gain consists of sensible heat due to radiation and convection from the body, and latent heat gain due to respiration and the evaporation of moisture from the skin. The proportion of sensible to latent heat emitted depends upon the age and sex of the occupant, the degree of activity, and the internal thermal environmental conditions.

Table 7.2 gives the total heat gain from an average adult male at different degrees of activity.

Table 7.2 Heat emission from the human body

Degree of activity	Total heat gain (W)
At rest	115
Sedentary work	140
Walking slowly	160
Light manual work	235
Medium manual work	265
Heavy manual work	440

Lighting: This comprises a sensible heat gain to the room, and the quantity of heat emitted will depend upon the type of fitting installed and the extent and usage of the fitting. Where the light fitting is specially ventilated so that the heat may be extracted, the heat transmitted to the room may be reduced below the nominal rating of the fitting. For normal light fittings 1 watt of lighting contributes 1 watt of heating.

Electrical machinery: Where electric motors are installed in the room the heat gain to the room will depend upon the following:

(*a*) the efficiency of the motor;
(*b*) whether or not the machinery being driven is also in the room;
(*c*) the frequency with which the motors will be used.

Gas appliances: The heat gain into the room from gas appliances will depend upon the following:

(*a*) the heat input to the appliance;
(*b*) the location of the appliance;
(*c*) whether it is connected to a flue or is flueless;
(*d*) position of any flue.

Cooking: The manufacturers of cooking equipment should provide details of the heat given out from their appliances. A table is given in the CIBS *Guide Book A,* Section 7.

Air infiltration (external heat gain)

If the room is maintained at a positive pressure, there will be little air infiltration, since all losses would be outwards. Where air filtrates by natural uncontrolled movement of air due to opening of doors and minute leakage through window frames, an allowance of one-half air change per hour is usually suitable.

Example 7.11 .   *Calculate the casual heat gains to an office given the following factors:*

1. *People 16*
2. *Lighting 20 W/m^2*
3. *Floor area 80 m^2*
4. *Height of room 2.5 m*
5. *Outside air temperature 30°C*
6. *Inside air temperature 22°C*
7. *Business machine load 400 W*
8. *Density of air 1.2 kg/m^3*
9. *Specific heat capacity of air 1012 J/kg*
10. *Infiltration half air change per hour*

Internal heat gains

$$\begin{aligned} \text{from people} &= (16 \times 115) = 1840 \text{ W} \\ \text{from lighting} &= (20 \times 80) = 1600 \text{ W} \\ \text{from machine} &= = 400 \text{ W} \\ \text{Total} &= 3840 \text{ W} \end{aligned}$$

Due to infiltration

$$\begin{aligned} \text{gain} &= \text{mass} \times \text{specific heat capacity} \times \text{temperature rise} \\ \text{mass} &= \text{volume of room} \times \text{air change} \times \text{density} \\ &= 240 \times 0.5 \times 1.2 \\ &= 144 \text{ kg/h} \\ &= 0.04 \text{ kg/s} \\ \text{gain} &= 0.04 \times 1012 \times (30 - 22) \\ &= 323.84 \text{ W} \end{aligned}$$

Total casual heat gains = 3840 + 323.84
= 4163.84 W

Admittance method

During warm sunny months, windows facing in a southerly direction are subject to daily cyclic heat gains from solar radiation in addition to other gains previously described.

In order to ensure that the rooms do not become uncomfortably hot during the sunny months, i.e., that the maximum peak indoor temperature does not frequently exceed about 27°C, the CIBS *Guide Book A,* Section 8, describes a technique known as the admittance method, which enables the peak indoor environmental temperature to be assessed for any proposed building design and also gives curves of peak indoor temperature against window size for a simple design of a multi-storey building.

Routine calculations

Application of the technique requires the following data to be calculated in turn:

1. Mean heat gains from all sources.
2. Mean internal environmental temperature.
3. Swing (deviation) from mean to peak in heat gains from all sources.
4. Swing (deviation) from mean to peak internal environmental temperature.
5. From 2 and 4, the peak internal environmental temperature.

Definitions

Many of the values used for heat-loss calculations are used but some new terms are also involved, namely:

1. **Admittance factor** (*Y*)   The factor which gives its name to the procedure, which is the amount of energy entering the surface for each degree of temperature swing at the environmental point. It is the reciprocal of the thermal resistance or impedance of an element to cyclic heat flow from the environmental temperature point and has the same units as *U* value (W/m^2 °C).

For slabs less than 75 mm thick, the admittance approximates to the *U* value of the structure, while for thicknesses above 200 mm, admittances tend to be a constant value. In comparison with lightweight materials, dense constructions have higher admittances, i.e., they absorb more energy for a given temperature swing.

For multi-layer slabs, the admittance is determined primarily by the surface layer, so that a 300 mm slab with 25 mm of insulation on the surface would respond more as a lightweight material than a heavy one. Table 7.3 gives admittance factors for some common types of construction.

2. **Environmental temperature**   Its use is essential in the admittance method and its value takes into account both the air temperature and the mean radiant temperature, i.e.,

$$\text{environmental temperature} = \tfrac{2}{3} \text{ mean radiant temperature} + \tfrac{1}{3} \text{ air temperature.}$$

Table 7.3 Admittance factors

Construction	Admittance (Y W/m² °C)
External walls	
Brick solid	
105 mm unplastered	4.2
105 mm with 16 mm lightweight plaster	3.1
Brick with 20 mm cavity (unventilated)	
105 mm inner and outer leaves with	
16 mm dense plaster on inner leaf	4.3
Brick with 20 mm cavity (unventilated)	
105 mm inner and outer leaves with	
16 mm lightweight plaster	3.3
Internal walls (partitions)	
105 mm brick with 15 mm dense plaster	
on both sides	3.3
Floors and ceilings	
Suspended timber floor and plasterboard ceiling	
Floor	0.1
Ceiling	0.3
150 mm cast concrete with 50 mm screed	
Floor	5.6
Ceiling	5.6
Windows	
Single glazed (unshaded)	5.6
Double glazed (unshaded)	5.6

3. Decrement factor The ratio of the cyclic transmittance to the steady-state U value.

4. Mean solar heat gains A function of the mean incident radiation intensity as read from tables in the *Guide Book*.

5. Mean casual heat gains The mean heat gain from casual sources such as lighting, machinery, etc., is found by multiplying the individual items by their duration and averaging over the 24-hour cycle.

6. Peak indoor environmental temperature The peak indoor environmental temperature is found by adding the mean-to-swing to the mean, thus:

$$t''_{ei} = t'_{ei} + \tilde{t}_{ei}$$

where t''_{ei} = peak internal environmental temperature (°C)

t_{ei} = mean internal environmental temperature (°C)

$\tilde{t}_{ei}$ = swing in internal environmental temperature (°C)

Example 7.12. *Estimate the peak indoor environmental temperature likely to occur during a sunny period in July for the office shown in Fig. 7.10. The characteristics of the office are given in Table 7.4.*

Step 1: Mean heat gains. The solar gain is found from the following formula:

94

Table 7.4 Characteristics of an office for example 7.12

Item	Detail
Window	Single glazing with internal blinds
Floor	Wood block on concrete
Ceiling	Plastered concrete
Partitions	Plastered brickwork
Occupancy	10 persons for 8 hours (80 W each)
Lighting	30 W/m² of floor, 7.00 to 9.00 and 16.00 to 18.00
Classification	Heavyweight construction

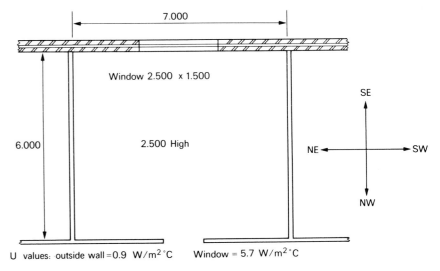

U values: outside wall = 0.9 W/m² °C   Window = 5.7 W/m² °C

Fig. 7.10 Plan of office — example 7.12

$$Q'_s = SI'A_g$$

where Q'_s = mean solar gain (W)

I' = mean solar intensity (W/m²)

S = solar gain factor

A_g = sunlit area of glazing

$$\therefore Q'_s = 0.46 \times 190 \, (2.5 \times 1.5)$$
$$= 327.75 \text{ W}$$

The mean casual heat gain is found from the following formula:

$$Q'_c = \frac{(q_{c1} \times t_1) + (q_{c2} \times t_2)}{24}$$

where Q'_c = mean casual gain (W)

q_{c1} and q_{c2} = instantaneous casual gains (W)

t_1 and t_2 = duration of individual casual gains in hours

$$Q'_c = \frac{(10 \times 80 \times 8) + (7 \times 6 \times 30 \times 4)}{24}$$

$$Q'_c = 476.7 \text{ W}$$

$$\therefore Q'_t = 327.75 + 476.7$$

$$= 804.5 \text{ W (approx.)}$$

Step 2: Mean internal environmental temperature. Assuming that the window is open during the day and closed at night, the ventilation loss is found from the following formula:

$$C_v = 0.33 \, N \, v$$

where N = rate of air change per hour

v = volume of room m^3

$$\therefore C_v = 0.33 \times 3 \times 7 \times 6 \times 2.5$$

$$= 103.95 \text{ W}$$

Fabric loss

$$\Sigma AU = 5.7 (2.5 \times 1.5) + 0.9 (17.5 - 3.75)$$

$$= (5.7 \times 3.75) + (0.9 \times 13.75)$$

$$= 33.75 \text{ W}$$

The mean indoor environmental temperature may now be found from the following equation:

$$Q'_t = (\Sigma AU + C_v)(t'_{ei} - t'_{ao})$$

where ΣAU = sum of products of areas of exposed surfaces and their U values (W/°C)

C_v = ventilation loss

t'_{ei} = mean internal environmental temperature °C

t'_{ao} = mean outdoor air temperature °C

The daily mean outdoor air temperature for July may be taken as 19 °C.

$$\therefore 804.5 = (33.75 + 103.95)(t'_{ei} - 17)$$

$$\therefore t'_{ei} = 24.84 \text{ °C (approx.)}$$

Step 3: Swing (mean to peak) in heat gain. The solar gain may be found from the following equation:

$$\tilde{Q}_s = S_a A_g (I_p - I')$$

where $\tilde{Q}_s$ = swing in effective heat gain due to solar radiation (W)

S_a = alternating solar gain factor

I_p = peak intensity of solar radiation (W/m^2)

Peak hour is 15.00 but allow for 2-hour time-lag, so peak intensity is at 13.00.

$$\tilde{Q}_s = 0.42 (2.5 \times 1.5) \times (280 - 190)$$

$$= 141.75 \text{ W}$$

The structural gain may be found from the following equation:

$$\tilde{Q}_f = f A U (t_{eo} - t'_{eo})$$

where $\tilde{Q}_f$ = swing in the effective heat input due to structural gain (W)

ϕ = time-lag in hours

f = decrement factor

t_{eo} = sol-air temperature at time of peak hours less time-lag (°C)

t'_{eo} = mean sol-air temperature (°C)

Assuming a time-lag of 6½ hours and a decrement factor of 0.4 (see Table 7.1):

$$\tilde{Q}_f = 0.4 \times 0.9 (17.5 - 3.75) \times (28.1 - 23.7)$$

$$= 4.95 \times 4.4$$

$$= 21.78 \text{ W}$$

(This amount is small enough to be ignored.)

Casual gain $\tilde{Q}_c = Q_c - Q'_c$

$$= (10 \times 80) - 476.7$$

$$= 323.3 \text{ W}$$

Gain air to air $\tilde{Q}_a = (\Sigma A_g U_g + C_v) t_{ao}$

where $\tilde{Q}_a$ = swing in effective heat input due to swing in outside temperature in (W)

$\Sigma A_g U_g$ = sum of products of areas of exposed glazing and their U values (W/°C)

$\tilde{t}_{ao}$ = swing in outside air temperature (°C)

$$\therefore \tilde{Q}_a = [(2.5 \times 1.5 \times 5.7) + 103.95] \, 6.5$$

$$= 814.6 \text{ W}$$

$$\therefore \tilde{Q}_t = 141.75 + 21.78 + 323.3 + 814.6$$

$$= 1301.43 \text{ W}$$

Step 4: Swing (mean-to-peak) in indoor environmental temperature using Table 7.3.

Floor	AY =	42	× 5.6 =	235.2	W/°C
Ceiling	AY =	42	× 5.6 =	235.2	W/°C
Window	AY =	3.75	× 5.6 =	21.0	W/°C
Outside wall	AY =	13.75	× 4.3 =	59.125	W/°C
Partitions	AY =	47.5	× 4.5 =	213.75	W/°C
			ΣAY =	764.275	W/°C

$$\tilde{Q}_t = (\Sigma AY + C_v) t_{ei}$$

where ΣAY = sum of products of all room surface, internal and external and their appropriate admittance values (W/°C)

$\tilde{t}_{ei}$ = swing in internal environmental temperature (°C)

$$\therefore 1301.43 = (764.275 + 103.95) \tilde{t}_{ei}$$

$$1301.43 = 868.225 \times \tilde{t}_{ei}$$

$$\therefore \tilde{t}_{ei} = 1.5\ ^{\circ}C$$

Step 5: Peak internal environmental temperature.

$$t''_{ei} = t'_{ei} + \tilde{t}_{ei}$$
$$= 24.84 + 1.5$$
$$= 26.34\ ^{\circ}C$$

Reasonable thermal conditions should therefore exist during the period of peak internal environmental temperature.

Solar heating in the UK

Solar collectors may be sited on the roof of a building and used to reduce the cost of providing hot water, or for space heating. They may also be sited on the ground close to swimming pools and used to heat the water.

A correctly installed solar water-heating system with the collectors facing south or west of south will intercept about 1000 kWh (3.6 GJ) of solar energy each year for each square metre of net panel area.

The overall efficiency of the system taken over the year is about 40 per cent and therefore the expected useful heat output from the system will be about 400 kWh/m^2 (1.44 GJ) per year.

For the average house with a system with a 5 m^2 of net panel area, about 2000 kWh (7.2 GJ) per year may be supplied to the hot-water system and this represents about 50 per cent of the total energy used annually for water heating.

On-peak electricity costs about 2.8p per kWh and the fuel-saving in this example will be about £50 per year. If the water is heated by off-peak electricity or gas, the fuel-saving will be between £20 and £40 per year because these fuels are cheaper than on-peak electricity. If 7 GJ of energy in the form of hot water could be provided for each 19 million households in the UK, the saving in primary energy would be about 4 per cent of the total consumption.

A normal domestic installation will require about 1 m^2 for every 50 litres of daily hot-water demand.

The Building Research Establishment have carried out a computer study of solar water heating and the following points summarise their findings:

1. The possible effects of a double-glazed panel is marginal and may, for small collector areas, be negative, the improved heat-retention properties of two sheets of glass not compensating for the poorer transmission of the incoming radiation.
2. Major changes in the angle of inclination of south-facing collectors produce only small variations in annual heat output. More heat is collected in winter when the panel is inclined at 60° to the horizontal. Collectors inclined at 40° to the horizontal provide the optimum yearly heat output.
3. Collectors facing west of south are preferable to those facing east of south: due south is not necessarily the optimum direction of solar collectors.
4. For a system with a 4 m^2 of net collection area, the optimum capacity of the solar tank is 200 litres.

Figure 7.11 shows a detail of a single-glazed solar collector and Fig 7.12 a detail of solar water-heating system which operates as follows:

1. The collector panels are heated by solar radiation which in turn heats the water inside them to a temperature of 50–60 °C. On a good summer's day it is possible for all the hot water required for an average house to be met by a system with a 4–5 m^2 collector area.
2. The pump is switched on by the control box when the temperature of the water at X exceeds that of Y by between 2–3 °C.
3. Hot water from the solar collector is pumped through the heat exchanger inside the solar cylinder which heats the water in the cylinder.
4. When the water is drawn off through the hot-water taps, the cold water from the cold-water storage cistern forces the hot water from the solar cylinder into the conventional cylinder and thus reduces or eliminates the heat required to raise the temperature of the water in the conventional cylinder to 60–70 °C. The water in the solar collector system has anti-freeze added to the water to prevent freezing. The expansion vessel takes up the expansion of the water in the solar collector system.

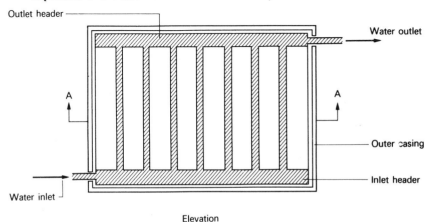

Elevation

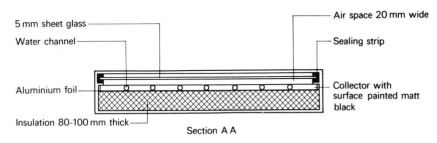

Fig. 7.11 Detail of solar collector

Solar space heating (see Fig. 7.13)

A solar-heated, timber-framed house having a floor area of 84 m^2 is planned to be built by the BRE. Incorporated into the roof will be 20 m^2 of water-heating panels which will supply heat either directly or via a heat pump, depending on the collection temperature, to a 35 m^3 insulated underground storage tank. Heat

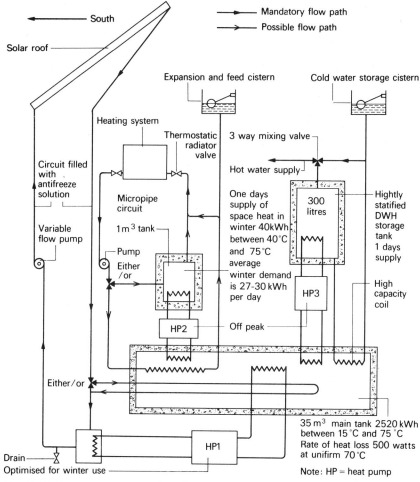

Fig. 7.13 Solar heated house (BRE)

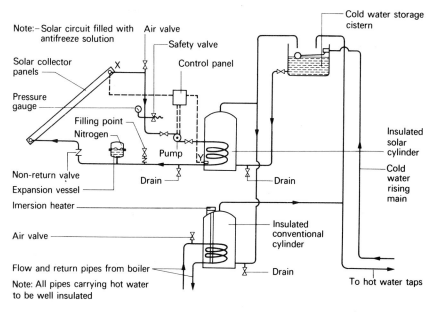

Fig. 7.12 Solar water-heating system

from this tank will be transferred during the night, via small heat pumps, to smaller heat accumulators of 1 m³ and 0.3 m³ which will supply the daily space- and water-heating requirements of the house.

The heat pumps will use electricity during the off-peak period at a reduced tariff.

Solar heating for swimming pools

One of the more viable applications of solar-energy utilisation in the UK is the heating of swimming pools, especially those that are only in use during the summer months.

Solar collectors having no glazing or insulation will generally be satisfactory for heating swimming pools because of the small temperature rises that will be required. The collectors should be located in a sheltered but not an over-shadowed position. An area of panel between half and three-quarters of the

surface area of the pool should give satisfactory results, especially if the surface of the pool is covered when not in use.

In most cases the pool water is circulated through the panels and this gives the most efficient results. Collectors should be made of material suitable for use with chlorinated water and both copper and some black plastics should be satisfactory.

Principles of operation of solar collector

A flat-plate collector consists of one or more cover plates of glass or trans-parent plastic. These materials transmit about 90 per cent of short-wave solar radiation, but allow less than 10 per cent of long-wave radiation to escape. This is often termed as the 'greenhouse effect' and ensures that solar heat is collected and not lost again by outward radiation. Coated-steel glass panels are claimed to increase the transmission of solar energy.

Questions

1. In respect to the determination of solar heat gains in buildings, define the terms altitude, azimuth, and declination.

2. Calculate the maximum and minimum altitudes of the sun in Liverpool, which lies on latitude 53.4° N.

Answers: maximum = 60.1°; minimum = 13.1°

3. Calculate the altitude of the sun for London at 16.00 hours sun time on 21 June, given that the latitude for London is 51.5° N.

Answer: 36.5° (approx.)

4. Calculate the total heating effect of the sun on a vertical window which faces south, given the following data:

time of day	noon
window size	2 m × 1.5 m
latitude	55.4° N
day and month	21 July
intensity of direct radiation normal to sun	800 W/m^2

Answer: 1380.48 W

5. Calculate the total heating effect of the sun on a horizontal flat roof given the following data:

time of day	noon
size of roof	8 m × 6 m
latitude	51.5° N
day and month	21 June
intensity of direct radiation normal to sun	850 W/m^2

Answer: 36 022.32 W

6. A horizontal glass skylight measures 3 m × 2 m. From the following data calculate the heat transmitted to the room by direct solar radiation.

time of day	noon
latitude	53.4° N
day and month	21 June
intensity of direct radiation normal to sun	760 W/m^2
glass transmission factor	0.43
glass absorption factor	0.42
percentage of absorbed radiation transmitted to room	66 per cent

Answer: 2795.56 W (approx.)

7. Calculate the total heating effect of the sun on a vertical window which faces southwest given the following data:

time of day	08.00 hours
altitude of sun	36°
latitude	51.7° N
wall-solar azimuth angle	62° W of S
intensity of direct radiation normal to sun	750 W/m^2
window size	2.5 m × 1.5 m

Answer: 1068.26 W (approx.)

8. A window facing directly into the sun measures 2 m long × 1 m high. If the altitude angle of the sun is found to be 40.5° and the solar radiation intensity 800 W/m^2, calculate the heating effect of the unshaded area of glass when the depth of recess is 150 mm.

Answer: 1395.2 W (approx.)

9. Define the term sol-air temperature, and describe how its value may be used to find the heat gains through a building structure.

10. State the sources of heat gains in a room other than those from solar radiation.

11. Explain the principle in which a flat-plate solar collector operates, and describe briefly the use of solar energy for hot-water supply and space heating.

Chapter 8

Heat pump and degree days

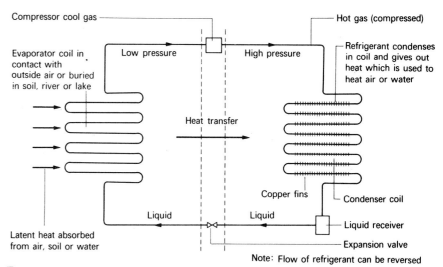

Fig. 8.1 The heat pump

The heat pump (see Fig. 8.1)

The heat pump is a device which extracts thermal energy from a low-temperature source and upgrades it to a higher temperature so that it may be used for space- or water heating. The principle of operation is similar to a refrigerator, but instead of wasting the heat given out in the condenser, this heat is utilised in the heat pump.

The low-temperature source in the system may be from air, soil, or water, which surrounds the evaporator. The main advantage of the heat pump is that it always provides more energy for heating than the energy used for driving it. The device may have a reversing valve so that the evaporator (which extracts heat) and the condenser (which gives out heat) may be interchanged. With a reversing valve, the heat pump may be used for heating in winter and cooling in summer.

Coefficient of performance (COP)

The theoretical maximum coefficient of performance can be expressed as follows:

$$\text{COP (max)} = \frac{t_c}{t_c - t_e}$$

where t_c = condenser temperature in degrees Kelvin

t_e = evaporator temperature in degrees Kelvin

It will be clear from the above equation that, in common with all vapour refrigeration systems, the heat pump operates to a greater advantage when working with a low condenser temperature and a higher evaporation temperature.

Example 8.1. *Calculate the theoretical coefficient of performance of a heat pump when the evaporator and condenser temperatures are −1°C and 50°C respectively.*

condenser temperature (absolute) = 50 + 273 = 323 K
evaporator temperature (absolute) = −1 + 273 = 272 K

$$\text{COP (max)} = \frac{t_c}{t_c - t_e}$$

$$= \frac{323}{323 - 272}$$

$$= 6.33$$

If the evaporator temperature is increased, the COP will also be increased; therefore in the example 8.1, if the evaporator temperature is increased to 10°C the theoretical COP will be:

evaporator temperature (absolute) = 10 + 273 = 283 K

$$\therefore \text{COP (max)} = \frac{323}{323 - 283}$$

$$= 8 \text{ (approx.)}$$

Theoretically, this would mean that for every 1 kW used to operate the compressor, the amount of heat received at the condenser would be 6.33 kW or 8 kW. In practice this does not happen since:

1. The actual working temperature difference between the evaporator and condenser temperatures is modified, due to the heat losses from the pipes.

2. The ideal vapour-pressure cycle requires stages of constant temperature condensation and expansion which cannot be achieved in practice.
3. The power absorbed by the fan to force the air over the condenser surfaces represents about 10 per cent of the total energy required to sustain the process.
4. Frictional losses in the compressor account for 10–15 per cent of the power required to operate the compressor.
5. The COP falls as the outside temperature falls. In practice, a realistic COP averaged over the UK heating season would be approximately 2.5, so that for every 1 kW of power consumed, 2.5 kW of useful energy would be available.

Solar collector combined with the heat pump

As mentioned earlier, the efficiency of the heat pump falls as the ambient temperature of the evaporator falls; therefore, if the temperature of air supplied to the evaporator is increased, the heat pump gives a corresponding increase in efficiency. This can be achieved by incorporating a solar collector in the system.

The solar collector consists of an air duct under the roof cladding; if single glazing is used instead of the normal roof cladding, the efficiency is increased, but at a higher capital cost.

An average increase in ambient temperature of the evaporator of $2.5\,^{\circ}\mathrm{C}$ during the heating season is possible, which increases the heat-pump efficiency by up to 10 per cent. Figure 8.2 shows the method of using a heat pump with a solar collector.

Actual coefficient of performance

If the actual COP of the heat pump can be kept above 2.8 it is more economical to burn primary fuels at the power station to generate electricity to run the heat pump, than to burn the primary fuel directly in the building. The actual COP is defined by:

$$\mathrm{COP_{act}} = \frac{\text{condenser heat (W)}}{\text{absorbed compressor power + heat losses (W)}}$$

Degree days

The generally accepted definition of a degree day is the daily difference in temperature in degrees Celsius between a base temperature of $15.5\,^{\circ}\mathrm{C}$ and the 24-hour mean outside temperature (when the outside temperature falls below $15.5\,^{\circ}\mathrm{C}$).

For most buildings, no heating will be required when the outside temperature is $15.5\,^{\circ}\mathrm{C}$ and due to internal heat gains from lighting, people, etc., the internal air temperature on average will be about $3\,^{\circ}\mathrm{C}$ above the external air temperature.

Uses of degree days

The degree day may be totalled for a month and these totals used to compare the monthly changes in the weather factor, or be added together for the heating

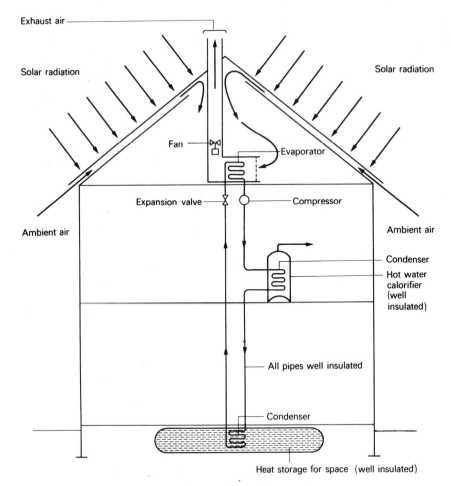

Fig. 8.2 Heat pump combined with solar heating

season and thus enable the severity and duration of the winter to be compared from year to year and from place to place. Table 8.1 gives a list of monthly and yearly degree days for several areas: a full list may be obtained from the Meteorological Office.

The monthly totals may be used to check the efficiency of a heating system, by comparing the fuel used during each month, during the heating season, with the fuel used over a previous month having the same number of degree days.

Example 8.2. *Calculate the degree day over 24 hours when the external maximum and minimum temperatures are $10\,^{\circ}\mathrm{C}$ and $5\,^{\circ}\mathrm{C}$ respectively.*

When the daily maximum and minimum external temperatures are both below $15.5\,^{\circ}\mathrm{C}$ the accumulated temperature in degree days may be found from the following formula:

$$\text{degree day} = 15.5 - \frac{(t_{\max} + t_{\min})}{2}$$

Table 8.1 Degree days for 12 months (base temperature 15.5°C)

	Sep	Oct	Nov	Dec	Jan	Feb	Mar	Apl	May	Jun	Jul	Aug	Total
Thames Valley	51	120	262	397	373	252	238	221	134	81	25	24	2178
South Eastern	77	139	296	421	387	272	261	258	163	102	42	42	2460
Southern	81	138	268	402	370	252	245	242	153	103	36	41	2331
South Western	67	127	218	334	322	232	224	228	139	86	21	34	2032
Severn Valley	72	151	263	411	377	254	244	227	152	100	27	33	2310
Midland	100	175	302	437	414	309	281	258	182	122	48	57	2685
West Pennines	89	153	284	417	395	300	262	254	156	101	39	40	2490
North Western	99	162	291	441	419	321	271	257	185	106	48	63	2663
Borders	114	170	296	418	412	334	295	262	224	149	77	72	2823
North Eastern	99	153	322	438	417	324	280	249	184	118	53	54	2705
East Pennines	88	158	304	425	414	318	278	245	166	106	47	44	2593
East Anglia	66	142	281	422	403	289	267	250	164	105	42	42	2473
West Scotland	95	186	309	456	429	344	284	258	187	103	49	65	2765
East Scotland	108	189	319	434	430	356	300	266	221	219	67	80	2899
North East Scotland	120	200	330	455	425	380	317	287	216	148	74	91	3043
Wales	81	169	242	368	362	273	257	255	172	120	41	47	2387
Northern Ireland	100	197	301	416	409	326	278	262	190	115	45	65	2704

$$= 15.5 - \frac{(10 + 5)}{2}$$

$$= 8$$

Example 8.3. *Calculate the degree day over 24 hours when the external maximum and minimum temperatures are 18°C and 8°C respectively.*

When the daily maximum temperature is above 15.5°C but by a lesser extent than the daily minimum temperature is below 15.5°C, the accumulated temperature in degree days may be found from the following formula:

$$\text{degree day} = \frac{1}{2}(15.5 - t_{min}) - \frac{1}{4}(t_{max} - 15.5)$$

$$= \frac{1}{2}(15.5 - 8) - \frac{1}{4}(18 - 15.5)$$

$$= 3.025$$

Example 8.4. *Calculate the degree day over 24 hours when the external maximum and minimum temperatures are 20°C and 14°C respectively.*

When the daily maximum external temperature is above 15.5°C but by a greater amount than the daily minimum temperature is below 15.5°C, the accumulated temperature in degree days may be found from the following formula:

$$\text{degree day} = \frac{1}{4}(15.5 - t_{min})$$

$$= \frac{1}{4}(15.5 - 14)$$

$$= 0.375$$

Example 8.5. *A building uses 3000 kg of oil during a winter month having 380 degree days. During a previous month having the same number of degree days, the same building used 2900 kg of oil. Calculate the loss in efficiency of the heating system when compared to the previous month.*

$$\text{efficiency compared to the previous month} = \frac{2900}{3000} \times \frac{100}{1} = 96.66 \text{ per cent}$$

$$\text{loss in efficiency} = 100 - 96.66 = 3.34 \text{ per cent}$$

Example 8.6. *A building uses 5000 kg of oil during a winter month having 340 degree days. During a previous month having 320 degree days, the same building used 5200 kg of oil. Calculate the loss in efficiency.*

$$\text{oil used during current month per degree day} = \frac{5000}{340} = 14.7 \text{ kg}$$

$$\text{oil used during previous month per degree day} = \frac{5200}{320} = 16.25 \text{ kg}$$

$$\text{increase in oil used per degree day} = 16.25 - 14.7 = 1.55 \text{ kg}$$

$$\therefore \text{loss in efficiency} = \frac{1.55}{14.7} \times \frac{100}{1} = 10.5 \text{ per cent}$$

Note: Fuel consumption, however, is not solely related to the external temperature. Other factors, such as wind strength, humidity, solar radiation, cloud, or doors and windows left open, may be very significant. The use of degree days for comparisons of fuel consumed should not therefore be regarded as leading to highly accurate results.

The present official base temperature of 15.5°C has been quoted and used in the calculations; it would, however, be reasonable to assume that the future base temperature might be 'rounded off' to 15°C or 16°C.

District heating

Degree days may also be used as a basis of comparison of the power used for district heating schemes in various parts of the country.

Example 8.7. *A district heating scheme of 700 dwellings in the South West required a boiler power of 9.4 MW. Calculate the boiler power required for a similar scheme in the North East. (Use figures from Table 8.1.)*

degree days per annum in South West = 2032
degree days per annum in North East = 2705

$$\text{boiler power required for North East} = \frac{2705}{2032} \times \frac{9.4}{1} = 12.5 \text{ MW (approx.)}$$

$$\text{difference} = 12.5 - 9.4 = 3.1 \text{ MW}$$

$$\text{percentage increase} = \frac{3.1}{9.4} \times \frac{100}{1} = 33 \text{ per cent (approx.)}$$

It would, if fuel costs are the same, cost about 33 per cent more for heating and hot-water supply in the North East than the South West. Degree days, however, vary from year to year and a more accurate comparison could be obtained by using the average degree days over, say, 10 years.

Questions

1. Draw a sketch of the heat pump and describe its operation.

2. Define the theoretical and actual coefficient of performance of the heat pump.

3. Calculate the theoretical coefficient of performance of a heat pump when the evaporator and condenser temperatures are 15 °C and 70 °C respectively.

Answer: 6.236

4. Define the term 'degree days' and explain how monthly totals of degree days may be used to check the efficiency of a heating system.

5. Calculate the degree day over 24 hours when the external maximum and minimum temperatures are 8 °C and 2 °C.

Answer: 3

6. A building uses 2000 kg of oil during a winter month having 410 degree days. During a previous month having the same number of degree days, the same building used 2200 kg of oil. Calculate the increase in efficiency of the heating system when compared with the previous month.

Answer: 9.1 per cent

7. State the other factors, besides external temperatures, that must be taken into account when comparing the efficiency of a heating system.

8. Explain how degree days may be used to compare the weather factors from year to year, and from place to place.

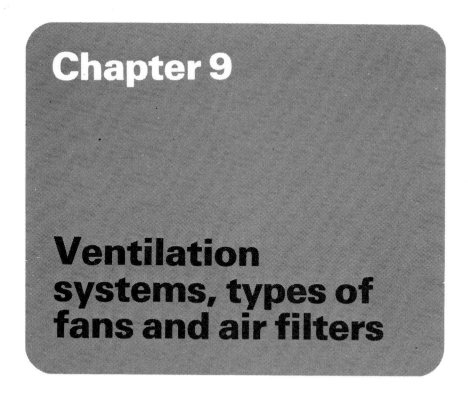

Chapter 9

Ventilation systems, types of fans and air filters

Air filters

The purpose of an air filter is to free the air of as much of the airborne contaminants as is practicable. Filters will justify their cost by a reduction in the cost of cleaning and decorating the building and the protection of the heating or air conditioning equipment. Filters are also required for various dust-free industrial processes. The main types of filters are:

1. *Dry:* in which the contaminants are collected in the filter medium.
2. *Viscous or Impingement:* in which the contaminants adhere to a special type of oil.
3. *Electrostatic:* in which the contaminants are positively charged with electricity and collected on negative earthed plates.

Dry filters

These use materials such as cotton wool, glass fibre, cotton fabric, treated paper, foamed polyurethane as the cleaning medium. The efficiency of the filter depends largely upon the area of medium offered to the air stream, and for this reason the filter can be arranged in a V formation which increases the area.

Figures 9.1, 9.2 and 9.3 show the arrangements of V type dry filter.

After a period of use (depending upon the atmospheric pollution) the contaminants retained by the filter will increase and this will increase the resistance to the flow of air through the filter.

In the case of an automatic roller type filter, when the filter is dirty, a pressure switch will switch on an electric motor which will turn the dirty spool and allow clean fabric to enter the filter chamber.

Figure 9.4 shows sections of two types of automatic roller type filter. Type A does not have as much filter area as type B.

Figure 9.5 shows a view of an automatic roll type filter.

With other types of dry filters an electrically operated pressure switch may be installed, which switches on a warning light to draw attention to a dirty filter. One type of dry filter is made up of a square cell of sizes from 254 by 254 mm to 600 by 600 mm. The filter medium, which is either 25 or 50 mm thick, is held in a metal or cardboard frame.

The cheaper type may be thrown away when dirty and the more expensive, vacuum-cleaned two or three times before being discarded.

Figure 9.6 shows a view of a disposable, or what is often called a 'throw-away dry cell type filter'.

Absolute filters

These are of dry fabric type and are very efficient in moving even the smallest particle from the air. This high performance is obtained by close packing of a very large number of small diameter fibres, but this unfortunately results in a high resistance to air passing through the filter.

Viscous filters

These have a large dust-holding capacity and are therefore often used in industrial areas where there is a high degree of atmospheric pollution. The filter medium is coated with a non-inflammable, non-toxic and odourless oil, which the contaminants adhere to as they pass through the filter.

There are two types of viscous filters:

1. *Cell type:* which consists of a metal frame into which wire mesh, industrial metal swarf, metal stampings or a combination of these materials are inserted and coated with the special oil. The cells are placed across the air stream in a V formation, similar to the dry cell type filter (shown in Fig. 9.1) which will allow the maximum area of the filter to be in contact with the air flow. When dirty, the cells are removed, washed in hot water, allowed to dry and re-coated with clean oil for further use.

Figure 9.7 shows a view of a cell type viscous filter.

2. *Automatic types:* one type uses a moving curtain, consisting of filter plates hung from a pair of chains. The chains are mounted on sprockets located at the top and bottom of the filter housing, so that the filter plates can be moved as a continuous curtain, up one side and down the other side of the sprockets. The arrangement is such that at the bottom the filter plates pass through a bath of special oil, which cleans the dirty oil from the filter plates and re-coates them with cleaner oil.

Figure 9.8 shows a vertical section of an automatic revolving type viscous filter. An electric motor is used to turn the sprockets and this may be arranged to move the curtain continuously, or periodically, depending upon the degree of atmospheric pollution. Another type has closely spaced corrugated metal plates, which are continuously coated by oil from a sparge pipe at the top of the filter.

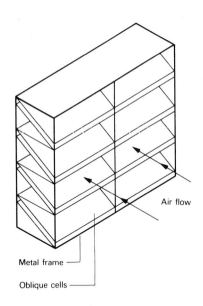

Fig. 9.1 Dry cell filter arranged in vee formation

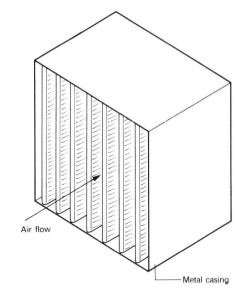

Fig. 9.2 Vee type dry fabric filter (plan)

Fig. 9.3 View of vee fabric filter

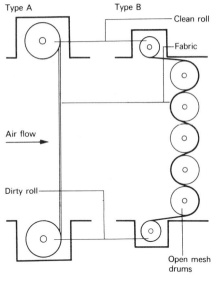

Fig. 9.4 Vertical sections of automatic roll type filter

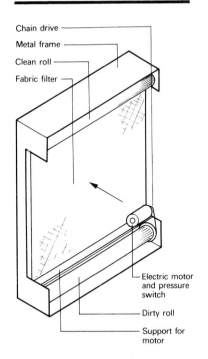

Fig. 9.5 View of automatic roll type filter

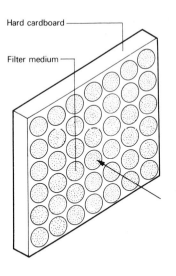

Fig. 9.6 Throw-away dry cell type filter

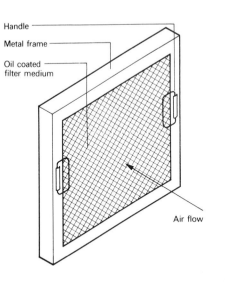

Fig. 9.7 Cell type viscous filter

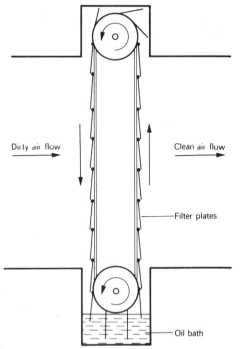

Fig. 9.8 Automatic revolving viscous

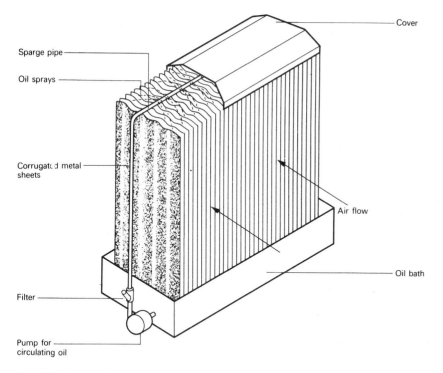

Fig. 9.9 Automatic spray tube viscous filter

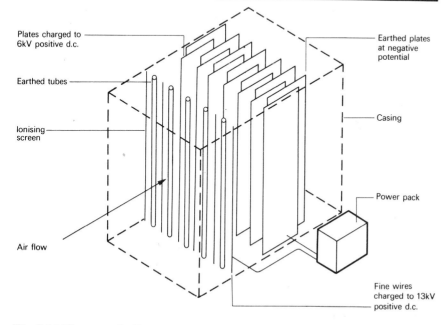

Fig. 9.10 Electrostatic filter

The air passing through the plates has to take a tortuous route and, in so doing, the dust particles in the air adhere to the oil which is washed down to an oil bath. The oil is pumped from this bath through a filter to the sparge pipe, where it is discharged over the filer plates thus recommencing the cycle.

Figure 9.9 shows a view of a spray type viscous filter.

Electrostatic filters

These types of filters have three main components: ioniser, metal collector and electrostatic power pack. The various air contaminants are given a positive electrostatic charge by an ioniser screen which is the first part of the filter. The screen consists of a series of fine wires possessing an electrostatic charge produced by a direct current potential of 13 kV. The wires are spaced alternatively with rods or tubes, which are at earth potential.

The air containing these positively charged contaminants then passes through a metal collector, which consists of a series of parallel plates about 6 mm apart, arranged alternatively so that one plate, which is earthed, is next to a plate which is charged with a positive direct current potential of 6 kV. The positively charged air contaminants passing through the collector are repelled by the plates of similar polarity (which are positive) and are attracted by the negative earthed plates, which are usually coated with a water-soluble liquid to permit easy cleaning. When the plates require cleaning, they are hosed down with warm water and it is sometimes found convenient to install sparge pipes above the plates for this cleaning process.

With some types of electrostatic filters the plates may be removed and immersed in a wash bath of warm water. Access to the filter is through air-tight doors with safety locks, so that it is impossible to open the access door without first switching off the electricity supply from the power pack. The resistance to the air flow through the filter is very low, but some prefilteration is usually necessary.

Figure 9.10 shows details of an electrostatic filter with the essential equipment.

Activated carbon filters

This is not a filter medium, but is used to remove odours, fumes and cooking smells from either dry or humid air. Activated carbon granules are used, which have a high absorptive capacity and this property causes a gas, or a vapour, to adhere to their surfaces. The granules are located in position by a glass fabric on either face of the filter panel. To obtain a maximum filtering area, the panels may be arranged in a V formation, as previously described for the dry and viscous filter and fitted inside a corrosion-resisting frame. They are often fitted across the inlets to ductwork serving cookers and fish fryers.

Measurement of contaminants

These may be in the form of a solid, liquid, gas or organic, and their size is measured in micrometre (μm) across the diameter of the particle. There are 1 million μm in 1 m, and 1000 in 1 mm. The smallest particle discernible to the human eye is about 10 μm; the human hair has a diameter of about 100 μm.

The sizes of the various contaminants are shown in Table 9.1.

Table 9.1

Contaminants	Average diameter (μm)
Tobacco smoke	0.05—0.1
Industrial smoke	0.1—1.0
Mist and fog	2.5—40
Atmospheric dust	1.0—100
Bacteria	0.3—10
Pollen	20—50
Virus	$\geqslant$1.0

Filter efficiencies

There are two main tests to determine the efficiency of a filter:

1. The determination of methylene blue efficiency, which uses the staining effect as a criterion. This test is applicable to any type of filter. A solution of methylene blue dye in water in injected into the air stream before the filter is tested. When the solution passes through the duct, the water is evaporated, leaving behind the blue dye in the form of small solid particles. A sample of the air containing the particles is drawn off before and after passing through the filter by means of a vacuum pump. The vacuum pump contains white filter paper, which is stained by the blue particles when the air is drawn through it. The extent of this staining is measured by means of a photosensitive cell. The values obtained are used to find the filter efficiency. The test is a stringent one and gives very low efficiency values for many filters.

2. The determination of the dust-holding capacity and gravimetric efficiency. This test does not apply to filters that require the accurate weighing of the filter, before and after the collection of the dust. The dust-holding capacity is rated in g/m^2 of filter face at a specified air velocity and the rating is only valid if the dust-holding capacity quoted is the same dust as used for the efficiency test.

Filter efficiencies

Dry: from 50 to 95 per cent in the 0.1 to 5 μm range.
Viscous: from 85 to 95 per cent in the 5 μm range.
Electrostatic: up to 99 per cent in the 0.1 μm range.

Ventilation

Principles

The purpose of ventilation of buildings is to remove high concentrations of body odours, carbon dioxide and water vapours and also to remove dust, fumes and smoke (which may be toxic) and excess heat. The air in the room containing these contaminants is replaced by fresh air and this creates air movement inside a building, so that the occupants obtain a feeling of freshness without draughts. At one time the concentration of carbon dioxide was used as a criterion of good ventilation, but even in very badly ventilated rooms the carbon dioxide rarely rises to a harmful level. The absence of body odours, dust and fumes in the air is a better criterion of good ventilation; also if the air movement is too low the air in the room will feel 'stuffy'. An air velocity of between 0.15 and 0.5 m per second is acceptable to most people under normal circumstances and higher velocities may be used for heavy manual work; to prevent monotony, a variable air speed is preferable to a constant air speed. The design of a ventilation system must be considered with the design of the heating system and it may be necessary to comply with the various public health legislations; for example, Factories Act 1961, Offices, Shops and Railway Premises Act 1963. A minimum ventilation rate of 28 m^3 of fresh air per hour, per person, is also required by most licensing authorities for theatres, cinemas and dance halls.

Systems of ventilation

Ventilation can be achieved by either natural or mechanical means. Natural ventilation for its operation depends on one of the following; (*a*) wind pressure, (*b*) stack effect, (*c*) a combination of wind pressure and stack effect. Wind causes a positive pressure to act on the windward side of the building and a negative pressure to act on the leeward side. Figures 9.11, 9.12 and 9.13 show the wind pressure distribution diagrams for pitched and flat roofs.

The inlet openings in the room should be well distributed and should be located on the windward side near the bottom. The outlet openings should also be well distributed and located on the leeward side near the top; this will allow cross ventilation in the room. Stack effect is created by the difference in temperature between the air inside and the air outside a building; the warmer, less dense air inside is displaced by the cooler denser air from outside. Figures 9.14 and 9.15 show upward and cross ventilation created by stack effect and Fig. 9.16 shows the stack effect in a tall building. The higher the column or 'stack' of warm air the greater will be the air movement inside the building; therefore in order that the force created by stack effect may be operated to its maximum advantage, the vertical distance between inlet and outlet openings should be as great as possible. If a wind is present and the air inside a building is warmer than the air outside, ventilation will take place, by both wind pressure and stack effect. Figure 9.17 shows how ventilation takes place through a casement window.

Unfortunately, natural ventilation cannot ensure a specified air change, nor is it possible to filter the air before it enters the building, also if the air inside is at the same temperature as the air outside, and there is no wind, natural ventilation will be non-existent. For the removal of excessive heat from founderies, bakeries, welding and plastic moulding shops natural ventilation is usually very successful, providing that the temperature differential is about 10 °C.

Mechanical ventilation

These systems employ an electrically driven fan or fans to provide the necessary air movement; they have the advantage over natural ventilation in providing positive ventilation at all times, irrespective of outside conditions. They also ensure a specified air change and the air under fan pressure can be forced through filters. There are three types of mechanical ventilation systems, namely:

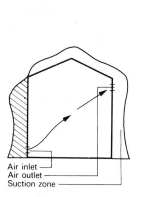

Air inlet
Air outlet
Suction zone

Fig. 9.11 Roof pitch 30

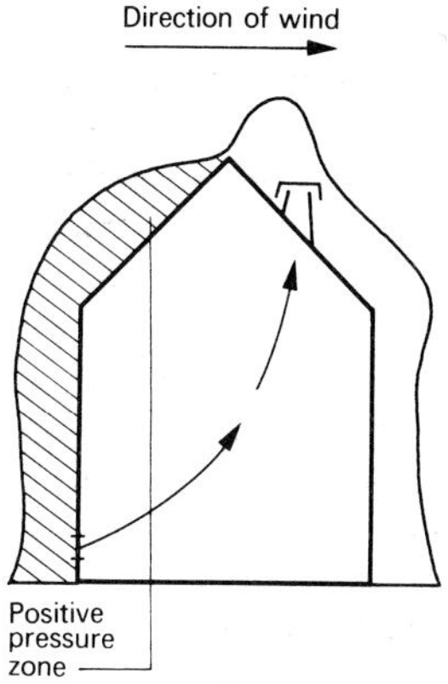

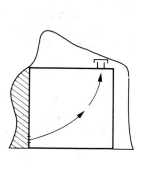

Direction of wind

Positive pressure zone

Fig. 9.12 Roof pitch over 30

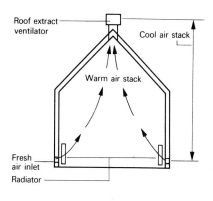

Fig. 9.13 Flat roof

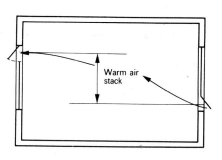

Roof extract ventilator
Cool air stack
Warm air stack
Fresh air inlet
Radiator

Fig. 9.14 Upward ventilation

Warm air stack

Fig. 9.15 Cross ventilation

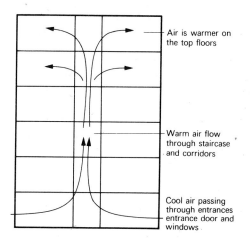

Air is warmer on the top floors

Warm air flow through staircase and corridors

Cool air passing through entrances entrance door and windows

Fig. 9.16 Stack effect in a tall building

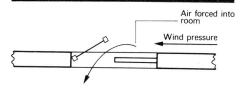

Air forced into room
Wind pressure

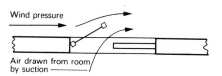

Wind pressure

Air drawn from room by suction

Fig. 9.17 Plan of casement window

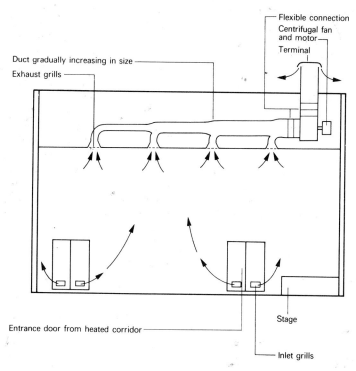

Duct gradually increasing in size
Exhaust grills
Flexible connection
Centrifugal fan and motor
Terminal
Entrance door from heated corridor
Stage
Inlet grills

Fig. 9.18 Ducted exhaust system for an assembly hall

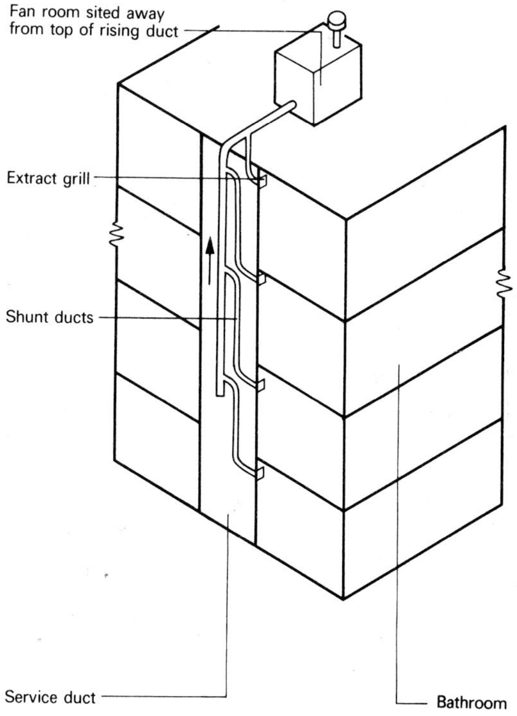

Fan room sited away from top of rising duct

Extract grill

Shunt ducts

Service duct

Bathroom

Fig 9.19 Common duct system for internal bathrooms and WC cubicles

1. *Natural inlet and mechanical extract* (exhaust system). This is the most common type of system and is used for kitchens, workshops, laboratories, internal sanitary apartments, garages and assembly halls. The fan creates a negative pressure on its inlet side, and this causes the air inside the room to move towards the fan, and the room air is displaced by the fresh air from outside the room.

Figure 9.18 shows a ducted exhaust system for an assembly hall. The air inlets can be placed behind radiators, so that the air is warmed before it enters the hall.

Internal bathrooms and W.C.s. Ventilation for these rooms should provide a minimum extract rate of 20 m³/h from a W.C. cubicle, or a bathroom without a W.C. and a minimum extract rate of 40 m³/h from a bathroom with a W.C. The ventilation system must be separate from any ventilation plant installed for any other purpose. In the common duct system the inlets from the bathrooms or W.C. compartments should preferably be connected to the main vertical duct by a shunt duct at least 1 m long. This shunt duct will offer better sound attenuation between the dwellings and also tends to prevent the spread of smoke and fumes in the event of a fire. The fans must be capable of extracting the total flow of air, plus an allowance on the fan static pressure, to counteract wind pressures. In order to keep the system operating in the event of failure of a single fan, it is recommended that two fans and motors are installed, with an automatic change-over damper. To replace air extracted from the rooms, air should be drawn from the entrance lobby through a wall grill, or a 19 mm gap left under the door of the room.

Figures 9.19 and 9.20 show the methods of ventilating internal bathrooms and W.C. cubicles in multi-storey blocks of flats. Figure 9.21 shows the method of duplicating the extract fans, including an automatic change-over damper.

2. *Mechanical inlet and natural extract.* It is essential with this system that the air is heated before it is forced into the building. The system may be used for boiler rooms, offices and certain types of factories. The air may be heated in a central plant and ducted to the various rooms, or a unit fan convector may be used. Figure 9.22 shows a ducted system for a multi-storey office block and Fig. 9.23 shows a system using a sill-level fan-heater unit.

3. *Mechanical inlet and extract.* This provides the best possible system of ventilation, but it is also the most expensive and is used for many types of buildings including cinemas, theatres, offices, lecture theatres, dance halls, restaurants, departmental stores and sports centres. The system is essential for operating theatres and sterilising rooms.

The air is normally filtered and provision is made for recirculation of the heated air which reduces fuel costs and, in order to save further fuel costs, the air may be extracted through the electric light fittings, which also increases the lighting efficiency about 14 per cent. Slight pressurisation of the air inside the building is achieved by using an extract fan smaller than the inlet fan; this requires the windows to be sealed and swing or revolving doors to be used. The slight internal air pressure and sealed windows prevent the entry of dust, draughts and noise.

Figure 9.24 shows a downward and upward air distribution system, with all the ductwork installed in the false ceiling or roof space. Figure 9.25 shows an upward air distribution system, having the heater unit in the basement. In both systems, the control dampers may be adjusted from a control panel, so that up to 75 per cent of the air may be recirculated thus saving a great deal of fuel costs.

Types of fans

There are three types of fans used for mechanical ventilation systems, namely: propeller, centrifugal and axial flow.

Propeller fans

These have two or more blades fixed at an angle to the hub. They develop a low pressure of only about 125 Pa (12.5 mm water gauge) and will therefore not force air through long lengths of ductwork. Their main purpose is for free air openings in walls or windows, but short lengths of ducting may be used, providing that the resistance through the duct is low. They can remove large volumes of air and their installation cost is low. A propeller fan having broad curved blades will move more air, and is quieter than a fan with narrow blades of the same diameter and running at the same speed. They operate at efficiencies of up to 50 per cent.

Figure 9.26 shows a four-bladed propeller fan and Fig. 9.27 shows the method of installing the fan at a wall opening, which is a common method used for ventilating workshops and kitchens.

Centrifugal fans

These consist of an impeller which revolves inside a casing shaped like a scroll. The impeller blades can be (*a*) forward or backward curved of either constant thickness or aerofoil section, (*b*) paddle blade. They can develop high pressure of up to about 760 Pa (76 mm water gauge) and are therefore used for forcing air through long lengths of ductwork, in both ventilating and air conditioning systems. The fans are generally quiet in operation but are bulky, the efficiencies range from 45 to 85 per cent, according to the type. The fan's output can be varied by different motors and speeds, or by being coupled to a bolt drive, which permits a change of pulley size to suit the fan's speed. The larger fans will deal with large volumes of air and are used extensively for large buildings. The inlet of the fan is at 90° to the outlet and this makes it sometimes difficult to install, unless the ductwork can also be turned through the same angle. Figure 9.28 shows the forward, backward and paddle blade impellers used with the centrifugal fan. The forward blade impeller is suitable for constant air resistance, the backward blade is suitable for variable air resistance and the paddle blade is suitable for air having a high level of suspended matter.

Figure 9.29 shows a view of a large centrifugal fan having a vertical outlet. Outlets from the fans can be vertical, up or down, horizontal, or at various angles to the horizontal.

Axial flow fans

These consist of an impeller with blades of aerofoil section, rotating inside a cylindrical casing. The air flows through the fans in a direction parallel to the

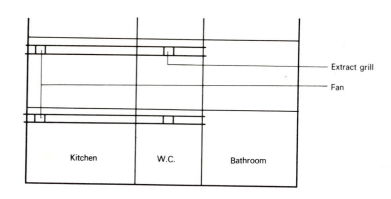

Fig. 9.20 Individual horizontal duct for internal bathrooms and W.C. cubicles

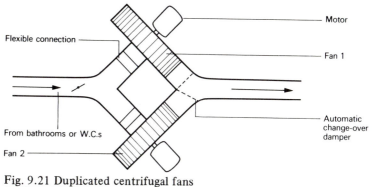

Fig. 9.21 Duplicated centrifugal fans

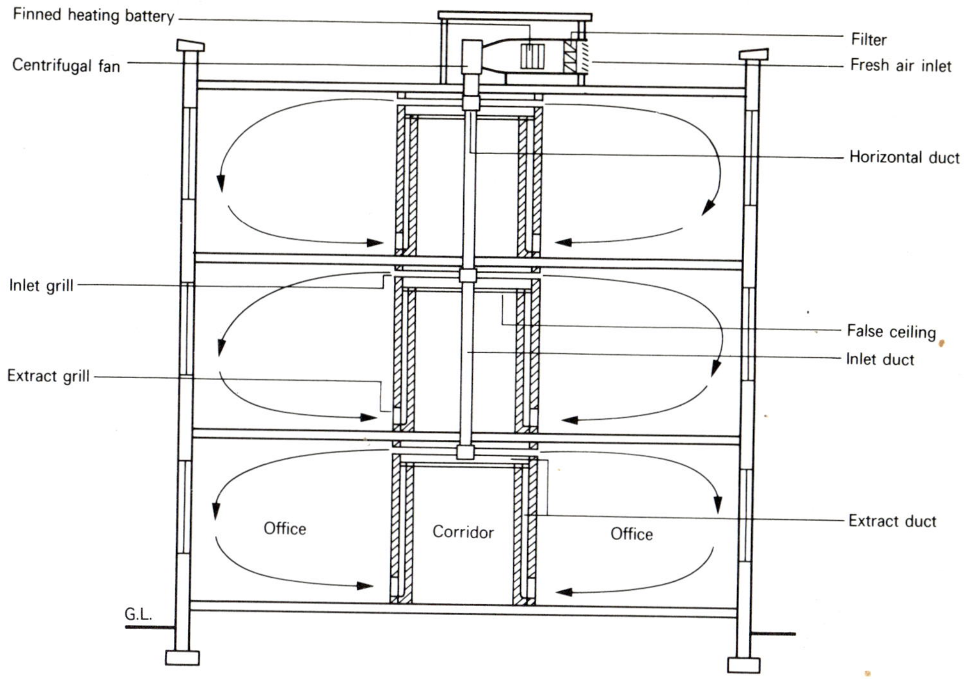

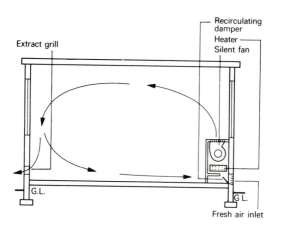

Fig. 9.23 System using sill-level fan-heater unit

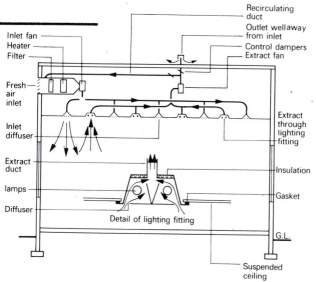

Fig. 9.24 Downward and upward system

Fig. 9.22 System for multi-storey office

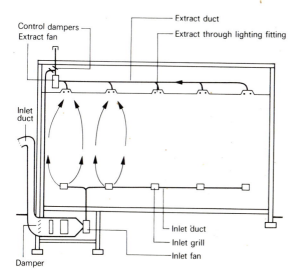

Fig. 9.25 Upward system

109

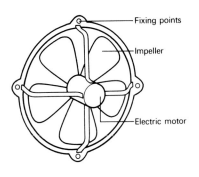

Fig. 9.26 Propeller fan

Fig 9.27 Propeller fan at wall opening

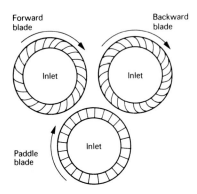

Fig. 9.28 Centrifugal fan impellers

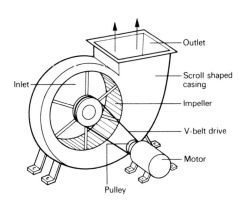

Fig. 9.29 View of centrifugal fan

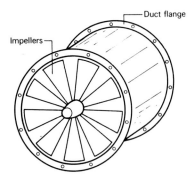

Fig. 9.30 View of axial flow fan

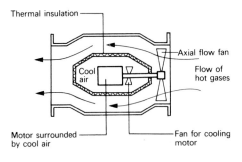

Fig. 9.31 Bifurcated fan

propeller shaft and can be installed without a fan base. The efficiencies of the fans range between 60 and 75 per cent and they can develop pressures of up to about 1500 Pa (150 mm water gauge). The fan can be driven directly by a motor mounted in the air stream, or by a belt drive from a motor mounted outside the duct. Like the centrifugal fan, it can be used to deal with large volumes of air, and is sometimes used in preference to the centrifugal fan, due to the ease of installation.

Figure 9.30 shows a view of an axial flow fan and Fig. 9.31 snows a bifurcated axial flow fan, which is used to extract hot air from founderies, or flue gases in the chimney.

Fan laws with constant air temperature

1. The volume of air delivered varies as the fan speed.
2. The pressure developed varies as the square of the fan speed.
3. The power absorbed varies as the cube of the fan speed.

Table 9.2 gives the minimum ventilation rates for different types of buildings: Code of Practice 352. Mechanical Ventilation and Air Conditioning of Buildings.

Table 9.2

Type of building	Air changes per hour	Type of building	Air changes per hour
Schools		Operating theatres	10
Classrooms	6	X-ray rooms	6
Assembly halls	3	Recovery rooms	3
Changing rooms	3	Entrances	3
Cloakrooms	3	Staircases	2
Dining rooms	3	Lavatories and bathrooms	2
Dormitories	3	Internal lavatories and	
Gymnasia	3	bathrooms	3
Common rooms	2	Duplicated fans required,	
Lavatories	2	and system distinct from	
Staff rooms	2	any other ventilating	
Laboratories	4	system	
Hospitals		Kitchens	20—40
Wards	3	Varying according to	
Dormitories	3	volume of air required	
Day rooms	3	through canopy	
Dining rooms	3	Laundries	10—20
Staff bedrooms	2	Boiler houses	10—15
Corridors	2	Smoking rooms	10—15

Note: For places of public entertainment such as cinemas, theatres, concert halls, assembly halls and dance halls the ventilation rate depends upon the Local Authorities Regulations, usually 28 m³ per hour, per person.

Table 9.3 gives the Institution of Heating and Ventilating Engineers, recommended outlet velocities from grills, for acceptable noise levels.

Table 9.3

Application	Maximum outlet velocity (m/s)
Libraries, sound studios, operating theatres	1.75—2.5
Churches, residences, hotel bedrooms, hospital rooms and wards, private offices	2.5—4.0
Banks, theatres, restaurants, classrooms, small shops, general offices, public buildings, ballrooms	4.0—5.0
Arenas, stores, industrial buildings, workshops	5.0—7.0

Chapter 10

Ventilation ducts, fan duty and characteristics

Ventilating ducts

The sizing of ducts to convey air from fans in a plant room to the various spaces inside a building relies upon empirical data for the evaluation of frictional losses. It cannot therefore be assumed that the calculated volumes of air will correspond with those actually delivered to the various discharge points, and for this reason dampers are incorporated inside the duct system. These dampers perform the same function as regulating valves in hot-water heating systems.

Flow of air in ducts

The Bernoulli theorem may be applied to air flow in ducts provided that an allowance is made for friction and separation. The theorem states that the total energy of each particle of a fluid is the same, provided that no energy enters or leaves the system at any point. If there is a loss of one type of energy, there must be a corresponding gain in another, or vice versa.

Figure 10.1 shows the total energies of air flowing inside a section of ductwork. The total energy possessed by the air is the sum of the potential, pressure and kinetic energies and this can be expressed as follows:

$$\text{total energy } H = z + \frac{P}{\rho g} + \frac{V^2}{2g} = \text{constant}$$

where z = potential energy due to the position of the particles or height above the datum (m)

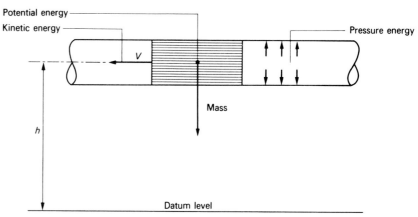

$\dfrac{P}{\rho g}$ = pressure energy due to the depth of the particles below the fluid (m)

$\dfrac{V^2}{2g}$ = kinetic energy due to the movement of particles of the fluid (m)

Fig. 10.1   Total energy of a moving fluid

Because the mass of air involved and its density are small, the value of the potential energy will also be small and for practical purposes is neglected. This reduces the equation to:

$$H = \frac{P}{\rho g} + \frac{V^2}{2g} = \text{constant}$$

Pressure distribution

In a closed system the total energy at two points must be equal, provided that there are no losses or gains.

$$\frac{P_1}{\rho g} + \frac{V_1^{\,2}}{2g} = \frac{P_2}{\rho g} + \frac{V_2^{\,2}}{2g}$$

If ideal conditions existed, i.e. no pressure loss due to friction, the distribution of pressure would be as shown in Fig. 10.2.

The loss of pressure energy due to friction between two points in a duct may be described in diagrammatic form (see Fig. 10.3).

The pressure energy $P/\rho g$ is referred to as the static head, or pressure of the system, and may be defined as the pressure acting equally in all directions. It is the pressure that tends either to burst the duct outwards when it is positive, or to collapse the duct inwards if it is negative.

The velocity energy $V^2/2g$ is referred to as the velocity head, or pressure of the system, and may be defined as the energy required either to accelerate the air from rest to a certain velocity, or to bring the air from a certain velocity to rest.

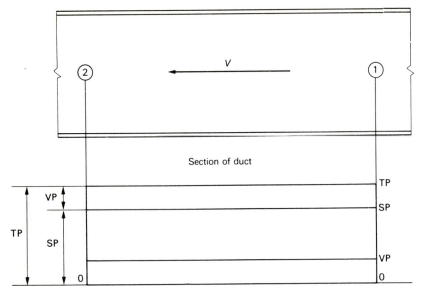

Section of duct

Pressure distribution
TP = Total pressure
SP = Static pressure
VP = Velocity pressure

Fig. 10.2 Total energy at two points in a ventilating duct for ideal conditions

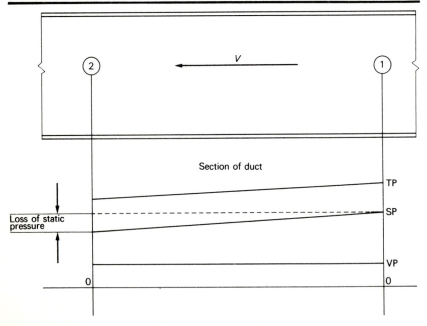

Section of duct

Loss of static pressure

TP = Total pressure   SP = Static pressure   VP = Velocity pressure

Fig. 10.3 Total energy at two points in a ventilating actual duct for conditions (i.e. pressure loss due to friction)

Measurement of pressure (see Fig. 10.4)

In order to measure the pressures inside a duct, a Pitot tube is used. One manometer will show the total pressure while the other will show the static pressure.

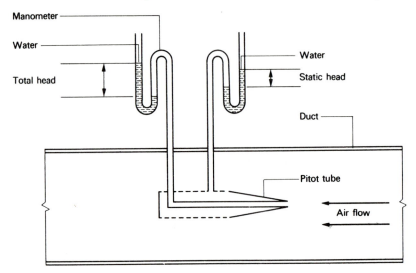

Fig. 10.4 Pitot tube

The velocity pressure may be found by deducting the static pressure from the total pressure, i.e.

$$\frac{\text{velocity pressure}}{\text{or head}} = \frac{\text{total pressure}}{\text{or head}} - \frac{\text{static pressure}}{\text{or head}}$$

Because the manometer-tube readings are very small, in order to obtain a higher degree of accuracy, an inclined manometer may be used with the Pitot tube.

The velocity pressure is always positive and since the total pressure is the sum of the static and velocity pressure, the total pressure may therefore be either positive or negative.

Velocity and velocity head (VH)

$$VH = \frac{V^2}{2g}$$

since the measurement is expressed in water gauge, and air is being considered:

$$VH = \frac{V^2}{2g} \times \frac{\rho_a}{\rho_w}$$

where VH = velocity head (m)
V = velocity of air flow (m/s)
ρ_a = density of air (1.2 kg/m^3)
ρ_w = density of water (998 kg/m^3)

$$\therefore VH = \frac{V^2}{2 \times 9.81} \times \frac{1.2}{998}$$

and $= V^2 \times 0.000\ 061\ 2$ (m) of water

In terms of mm of water,

$$VH = V^2 \times 0.0612$$

and

$$V = \frac{VH}{0.0612}$$

Velocity pressure (VP)

Manufacturers of fans often state the pressure developed by a fan in pascals instead of in millimetres water gauge. The pressure in pascals may be found as follows:

$$\text{pressure (Pa)} = \text{head of water (m)} \times 1000 \times 9.81$$
$$\therefore VP = V^2 \times 0.000\ 061\ 2 \times 1000 \times 9.81$$
$$= V^2 \times 0.6 \text{ Pa (approx.)}$$

This is the form given in the Chartered Institution of Building Services guide book and a table listing velocity pressure in pascals against velocity in m/s is produced.

Example 10.1. *Calculate the velocity pressure in pascals in a ventilating duct when the velocity of air is found to be 5 m/s.*

$$VP = 5^2 \times 0.6$$
$$= 15 \text{ Pa}$$

Example 10.2. *If the velocity pressure of air flowing in a duct is found to be 10 pascals, calculate the velocity of the air.*

$$VP = V^2 \times 0.6$$

$$V = \sqrt{\frac{VP}{0.6}}$$

$$V = 4 \text{ m/s (approx.)}$$

Example 10.3. *Calculate the velocity head in mm water gauge when the velocity of flow of air in a duct is 4 m/s.*

$$VH = 4^2 \times 0.0612$$
$$= 0.979 \text{ mm (approx.)}$$

Example 10.4. *Calculate the theoretical velocity of the flow of air in a duct when the velocity head is found to be 4 mm water gauge.*

$$V = \sqrt{\frac{4}{0.0612}}$$
$$= 8 \text{ m/s (approx.)}$$

Note: In practice, the velocity of air flow in a duct is often found by means of a velometer or anemometer.

Volume of air flow

The following formula may be used:

$$Q = VA$$
where Q = volume of air flow (m³/s)
V = velocity of air flow (m/s)
A = cross-sectional area of duct (m²)

Example 10.5. *The flow rate in a circular duct is to be determined by means of a Pitot tube. Determine the volume of flow in m³/s from the following data:*

$$\text{diameter of duct} = 600 \text{ mm}$$
$$\text{total head (TH)} = 30 \text{ mm water gauge}$$
$$\text{static head (SH)} = 18 \text{ mm water gauge}$$
$$\text{standard air density} = 1.20 \text{ kg/m}^3 \text{ at } 20^\circ C \;\longleftarrow$$

$$VH = TH - SH$$
$$= 30 - 18$$
$$= 12 \text{ mm water gauge}$$

$$V = \sqrt{\frac{12}{0.0612}}$$

$$V = 14 \text{ m/s}$$
$$Q = VA$$
$$Q = V \times 0.7854\ D^2$$
$$Q = 14 \times 0.7854 \times 0.600 \times 0.600$$
$$Q = 3.96 \text{ m}^3/s \text{ (approx.)}$$

Example 10.6. *A room measuring 20 m × 10 m × 3 m requires ventilating by means of a fan and ductwork to provide six air changes in the room. If the average velocity of air flow in the duct is to be 2 m/s, calculate the diameter of the main circular duct required for the room.*

$$Q = \frac{20 \times 10 \times 3 \times 6}{3600}$$

$$Q = 1 \text{ m}^3/s$$

$$Q = V \times 0.7854\ D^2$$

$$D = \sqrt{\frac{Q}{V \times 0.7854}}$$

$$D = \sqrt{\frac{1}{2 \times 0.7854}}$$

$$D = 0.798 \text{ m or } 798 \text{ mm}$$

The D'Arcy formula

It was shown in Chapter 3 that the loss of head due to friction for water flowing through a pipe was given in the following expression:

$$b = \frac{4fLV^2}{2gD}$$

The formula may be used for small circular ventilating ducts in the following terms:

$$b = \frac{4fLV^2}{2gD} \times \frac{\rho_a}{\rho_w}$$

where b = loss of static head due to friction (m)
f = coefficient of friction
L = length of duct (m)
V = velocity of flow (m/s)
D = diameter of duct (m)
ρ_a = density of air (1.2 kg/m^3)
ρ_w = density of water (998 kg/m^3)

$$\therefore b = \frac{4fLV^2}{2gD} \times 0.0012$$

The coefficient of friction depends upon the condition of the internal surface of the duct, and Reynolds number. For practical purposes, a value of between 0.005 and 0.007 may be used. The type of flow as well as the type of fluid affects the value, so that the D'Arcy formula acts as a guide only to the rate of frictional loss.

Example 10.7. *Calculate the static head lost due to friction in a 150 mm diameter ventilating duct 12 m long when the average velocity of air flow through the duct is 8 m/s.*

$$b = \frac{4 \times 0.007 \times 12 \times 8^2 \times 0.0012}{2 \times 9.81 \times 0.150}$$

h = 0.008 77 m or 877 mm

Rectangular ducts

In order to find the head lost due to friction, the D'Arcy formula may be modified as follows:

$$b = \frac{2(a+b)fLV^2}{2gab} \times \frac{\rho_a}{\rho_w}$$

$$h = \frac{2(a+b)fLV^2}{2gab} \times 0.0012$$

where a and b are the lengths of the sides of the rectangular duct in metres.

Example 10.8. *Calculate the static head lost due to friction in a rectangular ventilating duct having sides measuring 450 mm by 400 mm. The length of the duct is 30 m and the average velocity of air flowing through it 5 m/s.*

$$b = \frac{2(0.45 + 0.40) \times 0.007 \times 30 \times 5^2 \times 0.0012}{2 \times 9.81 \times 0.45 \times 0.40}$$

h = 0.003 m or 3 mm water gauge

Note: It is sometimes necessary to find the diameter of a circular duct having the same cross-sectional area as a rectangular duct, or vice versa. The diameter of the circular duct may be found from one of the following formulae, depending upon the problem in hand. Tables are also available for this purpose.

For equal velocity of flow:

$$d = \frac{2ab}{(a+b)}$$

For equal volume of flow:

$$d = 1.265 \times \left[\frac{(ab)^3}{a+b} \right]^{0.2}$$

where a and b are the lengths of the sides of the rectangular duct.

Duct sizing by use of chart (see Fig. 10.5)

In practical duct sizing, a chart is used which is derived from a more sophisticated formula than the D'Arcy, i.e. Colebrook-White equation.

Example 10.9. *Find by the use of the chart (see Fig. 10.5) the diameter of a circular duct 12 m long that will give a flow rate of 1 m^3/s when the velocity of flow is 8 m/s. Find the static head lost due to friction in mm water gauge.*

Figure 10.6 shows how the chart should be used and the following values are found:

1. diameter of duct = 400 mm
2. frictional loss per metre run = 0.24 mm
3. total frictional loss = 0.24 × 12
   = 2.88 mm

This may be used as a check against the D'Arcy formula previously explained:

$$b = \frac{4 \times 0.007 \times 12 \times 8^2 \times 0.0012}{2 \times 9.81 \times 0.400}$$

h = 0.003 287 m or 3.2 mm

The D'Arcy formula gives 0.32 mm higher frictional loss, which is negligible in this example.

Loss of head or pressure due to fittings

The resistance to the flow of air in ducts due to fittings such as bends, branches, dampers, etc., may be expressed in head of water and this value may then be converted into pressure in pascals, as described previously. The fundamental expression for the velocity head for the flow of water is given in the following expression:

$$b = \frac{V^2}{2g}$$

For the flow of air in ducts this becomes

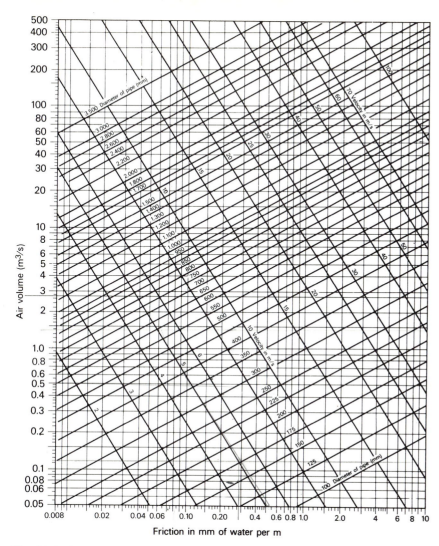

Fig. 10.5 Duct-sizing chart

$$b = \frac{V^2}{2g} \times \frac{\rho_a}{\rho_w}$$

and expressing the resistance of fittings in terms of velocity head, the following expression results:

$$b = k \quad \frac{V^2}{2g} \quad \frac{\rho_a}{\rho_w}$$

where b = velocity head loss due to fittings (m)
k = velocity head factor determined experimentally
V = velocity of air flow (m/s)

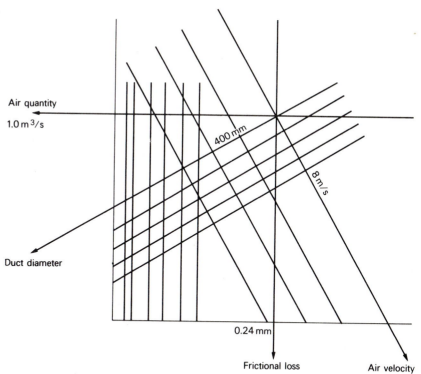

Fig. 10.6 Method of using duct-sizing chart for example 10.8

ρ_a = density of air (1.2 kg/m³ at 20 °C)
ρ_w = density of water (998 kg/m³)

Table 10.1 gives some general values for velocity-head factors for illustration purposes only. A full list is given in the CIBS guide book.

Table 10.1 Velocity-head loss factors (k)

Type of fitting	k **factor**
90° Rounded elbow	0.65
90° Square elbow	1.25
90° Bend ($R = 2D$)	0.10
Branch	0.6–1.3
Abrupt enlargement	0.30
Gradual enlargement	0.15
Abrupt reducing	0.30
Gradual reducing	0.04
Open damper	0.30
Wire mesh	0.40
Diffuser	0.60
Outlet	1.00

Example 10.10. *Determine the total loss of head in mm water gauge in a 300 mm diameter duct 15 m long, having three 90° bends when the velocity of air flow is 6 m/s.*

$$\text{velocity-head loss due to bends} = 3 \times 0.1 \left[\frac{6^2}{2 \times 9.81}\right] \times 0.0012$$

$$= 0.000\ 66 \text{ m}$$
$$= 0.66 \text{ mm}$$

Static-head loss due to straight duct, using the duct-sizing chart (Fig. 10.5) for 300 mm diameter and 6 m/s.

$$\text{for 1 m} = 0.19 \text{ mm}$$
$$\text{for 15 m} = 0.19 \times 15$$
$$= 2.85 \text{ mm}$$
$$\text{total head loss} = 0.66 + 2.85$$
$$= 3.51 \text{ mm}$$

Sizing of system of ductwork

Several methods may be used to determine the sizes of ventilating ducts to deliver or extract air from various spaces inside a building. In theory, duct sizing is very complex, but the following approximate methods are used in practice:

Equal-velocity method: The air velocity in the ductwork is decided by the designer and duct cross-sectional areas obtained from the volumes of flows required using the continuity equation. This method is generally used for simple systems without branches and for pneumatic conveying.

Velocity-reduction method: The air velocity in the first section of the main duct is chosen and the velocity reduced at each branch. Duct sizes are then found using the continuity equation.

Static-regain method: The size of a duct between two adjacent branches is chosen so that the pressure drop over the section is equal to the static regain at the previous branch. Therefore the static pressure at every branch is the same.

Constant pressure drop per unit length: The air velocity in the first section of the main duct is chosen and the size of the duct in this section found. Using the duct-sizing chart, the pressure drop per unit length of duct is obtained for this section and this value is then used for all the remaining sections of ductwork. This is a very convenient method of duct sizing and will be the one used in the following example.
 For low-velocity systems a total pressure drop of 1 Pa may be used.

Example 10.11. *Figure 10.7 shows an extract system of ductwork. Determine by use of the duct sizing chart (Fig. 10.5) the diameters of the ducts A, B and C assuming that the average velocity of air flow through duct A is to be 5 m/s.*

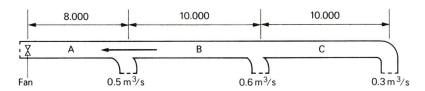

Fig. 10.7 System of ductwork

Duct A

$$\text{volume of flow} = 0.5 + 0.6 + 0.3$$
$$= 1.4 \text{ m}^3\text{/s}$$

From the duct-sizing chart a flow rate of 1.4 m³/s and an air velocity of 5 m/s gives a head loss per metre run of 0.059 mm water gauge and a 600 mm diameter duct.

Duct B

Using the same head loss per metre run as for duct A,

$$\text{flow rate} = 0.6 + 0.3$$
$$= 0.9 \text{ m}^3\text{/s}$$

From the duct-sizing chart a 500 mm diameter duct giving an air velocity of approximately 4.4 m/s is found to be suitable.

Duct C

Using the same head loss per metre run as for duct A,

$$\text{flow rate} = 0.3 \text{ m}^3\text{/s}$$

From the duct-sizing chart a 350 mm diameter duct giving an air velocity of approximately 3.4 m/s is found to be suitable.

Total head loss

The values of resistances through the various fittings and the lengths of ductwork can now be calculated basing the velocity head on the velocity through each section of ductwork. The total loss of pressure through the duct will be required when ordering the fan.

Duct A

$$\text{loss of head through straight duct} = 8 \times 0.059$$
$$= 0.422 \text{ mm}$$
$$= 0.000\ 422 \text{ m}$$

To this must be added the loss of head through the outlet, which from Table 10.1 has a *k* factor of 1.0.

$$\text{loss through outlet} = 1 \left[\frac{5^2}{2 \times 9.81}\right] \times 0.0012$$

$$= 0.0015 \text{ m}$$
$$\text{total loss} = 0.000\ 422 + 0.0015$$
$$= 0.001\ 922 \text{ m}$$
$$= 0.002 \text{ m (approx.)}$$

Duct B

loss of head through straight duct $= 10 \times 0.059$

$$= 0.59 \text{ mm}$$
$$= 0.000\ 59 \text{ m}$$

To this must be added resistances of two branches and two inlets.

branches (average k, 0.9) $= 2 \times 0.9 \left[\dfrac{4.4^2}{2 \times 9.81} \right] \times 0.0012$

$$= 2 \times 0.9 \times 0.987 \times 0.0012$$
$$= 0.002\ 13 \text{ m}$$

inlets (wire mesh k, 0.4) $= 2 \times 0.4 \left[\dfrac{4.4^2}{2 \times 9.81} \right] \times 0.0012$

$$= 2 \times 0.4 \times 0.987 \times 0.0012$$
$$= 0.000\ 947 \text{ m}$$

total loss $= 0.000\ 59 + 0.002\ 13 + 0.000\ 947$

$$= 0.003\ 667 \text{ m}$$
$$= 0.004 \text{ m (approx.)}$$

Duct C

loss of head through straight duct $= 10 \times 0.059$

$$= 0.59 \text{ mm}$$
$$= 0.000\ 59 \text{ m}$$

To this must be added one bend and one inlet.

bend (k, 0.1) $= 0.1 \left[\dfrac{3.4^2}{2 \times 9.81} \right] \times 0.0012$

$$= 0.1 \times 0.589 \times 0.0012$$
$$= 0.000\ 070\ 6 \text{ m}$$

inlet (k, 0.4) $= 0.4 \left[\dfrac{3.4^2}{2 \times 9.81} \right] \times 0.0012$

$$= 0.4 \times 0.589 \times 0.0012$$
$$= 0.000\ 287 \text{ m}$$

total loss $= 0.000\ 59 + 0.000\ 07 + 0.000\ 287$

$$= 0.000\ 947 \text{ m}$$

loss through duct system $= 0.002 + 0.004 + 0.0095$

$$= 0.0155 \text{ m}$$

Converting this loss of head to pressure (Pa)

pressure $= 0.0155 \times 1000 \times 9.81$

$$= 152 \text{ Pa}$$

Fan duty

A fan that will discharge 1.4 m³/s and develop a total negative pressure of

152 Pa would be suitable. Dampers placed in the inlet branches will enable the system to be balanced.

Fans

There are three main types of fans used for ventilation systems:

Propeller fan: This type is usually used at free opening in walls or windows and for other types of low pressure applications. It is not usually employed for ducted ventilation systems.

Axial-flow fan: This type of fan is designed for mounting inside a duct system and is suitable for moving air in complete systems of ductwork.

Centrifugal fan: The inlet of the fan is at 90° to the outlet and like the axial-flow fan is suitable for moving air in complete systems of ductwork.

Fan laws

The performance of a fan incorporated in a system of ventilation is governed by the following laws, providing the air density remains constant:

1. The discharge varies directly with the angular velocity of the impeller.
2. The pressure developed varies directly as the square of the angular velocity of the impeller.
3. The power absorbed varies as the cube of the angular velocity of the impeller.

The laws may be expressed as follows:

1. $$\dfrac{Q_2}{Q_1} = \dfrac{N_2}{N_1}$$

2. $$\dfrac{P_2}{P_1} = \dfrac{(N_2)^2}{(N_1)^2}$$

3. $$\dfrac{\text{power}_2}{\text{power}_1} = \dfrac{(N_2)^3}{(N_1)^3}$$

where $Q =$ volume of flow in m³/s

$N =$ revolutions of impeller per minute

$P =$ pressure in pascals

power $=$ power in watts or kilowatts

Example 10.12. *A fan absorbs 2.3 kW of power and discharges 2.5 m³/s when the impeller angular velocity is 1000 revolutions per minute. If the impeller angular velocity is increased to 1200 revolutions per minute, calculate the discharge in m³/s and the power absorbed for this new condition.*

$$\dfrac{Q_2}{Q_1} = \dfrac{N_2}{N_1}$$

$$\therefore Q_2 = \dfrac{Q_1 \times N_2}{N_1}$$

119

$$Q_2 = \frac{2.5 \times 1200}{1000}$$

$$= 3 \text{ m}^3/\text{s}$$

$$\frac{\text{power}_2}{\text{power}_1} = \frac{(N_2)^3}{(N_1)^3}$$

$$\therefore \text{power}_2 = \frac{W_1 \times (N_2)^3}{(N_1)^3}$$

$$= \frac{2.3 \times 1200^3}{1000^3}$$

$$= 3.9744 \text{ kW}$$

It will be observed from example 10.12 that for a 20 per cent increase in angular velocity there is a 72.8 per cent increase in power absorbed. The common method used to lower the air flow through a main duct is by closing a damper, and this can lead to a significant loss of power. An alternative method of lowering the air flow is by lowering the angular velocity of the impeller. This can be achieved by varying the frequency of the current input to the motor driving the fan.

Example 10.13. *A fan develops a static pressure of 200 Pa when the angular velocity of the impeller is 900 revolutions per minute. If the angular velocity of the impeller is increased to 1000 revolutions per minute, calculate the static pressure developed by the fan for the new condition.*

$$\frac{P_2}{P_1} = \frac{(N_2)^2}{(N_1)^2}$$

$$\therefore P_2 = \frac{P_1 \times (N_2)^2}{(N_1)^2}$$

$$= \frac{200 \times 1000^2}{900^2}$$

$$= 246.9$$
$$= 247 \text{ Pa (approx.)}$$

Efficiency of fans

The efficiency of a fan can be found by use of the following expressions:

$$\frac{\text{efficiency}}{\text{(per cent)}} = \frac{\text{fan total pressure} \times \text{volume of flow}}{\text{power absorbed (W)}} \times \frac{100}{1}$$

Example 10.14. *A fan absorbs 1.5 kW when discharging 2 m³/s of air and operating at a total pressure of 500 Pa. Calculate the percentage efficiency of the fan.*

$$\text{efficiency} = \frac{500 \times 2}{1500} \times \frac{100}{1}$$

$$= 66.667 \text{ per cent}$$

Change of air density

If the air density is changed, the following laws apply:

1. The volume of flow remains constant.
2. The pressure developed varies directly with the change in density.
3. The power absorbed varies directly with the change in density.

The laws may be expressed as follows:

1. $\quad Q_1 = Q_2$

2. $\quad \dfrac{P_2}{P_1} = \dfrac{\rho_2}{\rho_1}$

3. $\quad \dfrac{\text{power}_2}{\text{power}_1} = \dfrac{\rho_2}{\rho_1}$

where Q = volume of flow in m³/s
$\quad\quad P$ = pressure in pascals, which may be static, velocity or total
$\quad\quad \rho$ = density of air in kg/m³
power = power in watts or kilowatts

Example 10.15. *A fan develops a total pressure of 400 Pa when discharging air at a temperature of 20 °C. If the temperature of the air is lowered to 16 °C, calculate the pressure to be developed by the fan in order to discharge the same volume of air.*

density of air at 20 °C = 1.2 kg/m³
density of air at 16 °C = 1.22 kg/m³

$$\frac{P_2}{P_1} = \frac{\rho_2}{\rho_1}$$

$$\therefore P_2 = \frac{P_1 \times \rho_2}{\rho_1}$$

$$= \frac{400 \times 1.22}{1.2}$$

$$= 406.67 \text{ Pa}$$

Selection of a fan

In order to select a fan for a given duty, reference should be made to fan-performance graphs supplied by the manufacturer.

Figure 10.8 shows a typical performance graph for two fans and it will be seen that the following factors may be obtained from them:

1. pressure in pascals;
2. volume flow rate in m³/s;
3. input in watts.

Before using the graphs it is essential to draw a system characteristic curve and the intersection of this curve with the fan-performance curve will give the actual amount of discharge and the pressure developed by the fans.

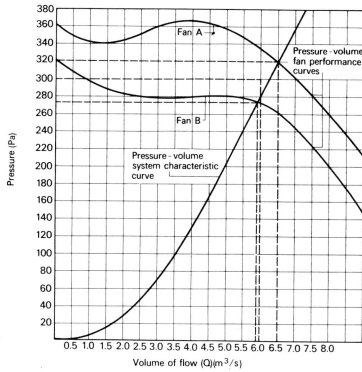

Note: Manufacturers of fans will supply graphs giving pressure, volume of air flow, power and efficiency

Fig. 10.8   Performance graph for fans

System characteristic curve

For a certain rate of flow of air through a system of ductwork a certain static pressure must be developed.

For any system of ductwork therefore, a characteristic curve may be drawn plotting the rate of flow against the static pressure to be developed in order to overcome the resistances.

Example 10.16.   *A ventilating system requires a fan to discharge 6 m³/s against a calculated resistance of 300 Pa pressure. Select from the graph in Fig. 10.8 either fan A or fan B.*

For a given system of ductwork the pressure loss may be found from the following expression:

pressure loss = pressure-loss coefficient × volume of flow²

$$P = k\,Q^2$$

$$\therefore k = \frac{P}{Q^2}$$

$$= \frac{300}{6^2}$$

$$= 8.333$$

In order to draw the system characteristic curve it is necessary to find the values of pressure corresponding to nominated values of flow. Using the flow rates between 0.5 and 6 m³/s :

1. $Q = 8.33 \times 0.5^2 = 2.08$ Pa
2. $Q = 8.33 \times 1.0^2 = 8.33$ Pa
3. $Q = 8.33 \times 1.5^2 = 18.74$ Pa
4. $Q = 8.33 \times 2.0^2 = 33.32$ Pa
5. $Q = 8.33 \times 2.5^2 = 52.06$ Pa
6. $Q = 8.33 \times 3.0^2 = 74.97$ Pa
7. $Q = 8.33 \times 3.5^2 = 102.04$ Pa
8. $Q = 8.33 \times 4.0^2 = 133.28$ Pa
9. $Q = 8.33 \times 4.5^2 = 168.68$ Pa
10. $Q = 8.33 \times 5.0^2 = 208.25$ Pa
11. $Q = 8.33 \times 5.5^2 = 251.98$ Pa
12. $Q = 8.33 \times 6.0^2 = 299.88$ Pa

By reference to Fig. 10.8, fan A will discharge 6.5 m³/s and develop a pressure of 320 Pa for the conditions given. Fan B would discharge 5.9 m³/s and develop a pressure of 272 Pa for the same conditions.

The designer has a choice between fan A and fan B, and fan A would probably be chosen. If it is required to save electrical power, however, fan B might be selected.

Limiting velocities in ducts

In order to reduce noise and power consumption, the velocity of air flowing through a duct must be kept within reasonable limits. Table 10.2 gives values of limiting velocities in ducts.

Table 10.2   Limiting velocities in ducts

Application	Velocity (m/s)	
	Commercial buildings	Industrial buildings
Outside air intake	5.0	7.5
Discharge to atmosphere	5.0	7.5
Main supply or extract duct	7.5	13.0
Terminal branch duct	3.5	5.0

Questions

1. Make a statement of Bernoulli's theorem as it applies to air flow in duct-work.

2. Define velocity, static and total pressure and describe how the Pitot tube may be used to measure the velocity, static and total pressures of air flowing through a duct.

3. Calculate the velocity pressure in pascals in a ventilating duct when the velocity of air is found to be 7 m/s.

Answer: 29.4 Pa

4. Calculate the velocity of air flowing in a duct when the velocity pressure is 25 Pa.

Answer: 6.455 m/s

5. Calculate the flow rate in m³/s through a 350 mm diameter ventilating duct when the total and static heads are 30 mm water gauge and 25 mm water gauge respectively. Assume standard air density of 1.20 kg/m³.

Answer: 0.87 m³/s (approx.)

6. A room measuring 15 m × 8 m × 3 m requires ventilating by means of a fan and ductwork to provide three air changes per hour. If the average velocity of air flow in the duct is to be 4 m/s, calculate the diameter of the main circular duct required for the room.

Answer: 309 mm

7. Calculate the static head lost due to friction in a 300 mm diameter ventilating duct 15 m long when the average velocity of air flow through the duct is 5 m/s. Use the following values:

(a) density of water   = 998 kg/m³
(b) density of air   = 1.2 kg/m³
(c) coefficient of friction = 0.005

Answer: 1.53 mm water gauge

8. Calculate the static head lost due to friction in a rectangular ventilating duct having sides 350 mm by 300 mm. The length of the duct is 12 m and the average velocity of air flowing through it 3 m/s. Use the values given in Question 7.

Answer: 0.8 mm water gauge

9. Find by use of the duct-sizing chart (Fig. 10.5) the diameter of a circular duct 50 m long that will give a flow rate of 9 m³/s when the velocity of flow is 5 m/s. Find the total static head lost due to friction in mm water gauge.

Answers: diameter 1.4 m; total static head lost 0.9 mm water gauge

10. Determine the total loss of head in mm water gauge in a 500 mm diameter duct 30 m long, having four 90° bends, when the velocity of air flow is 5 m/s.

Answer: 2.8 mm water gauge (approx.)

11. Figure 10.9 shows an inlet system of ductwork. Determine by use of the duct-sizing chart (Fig. 10.5) the diameters of the ducts A and B. It may be assumed that the average velocity of air flow through duct A is to be 6 m/s.

Answers: duct A 500 mm diameter; duct B 450 mm diameter

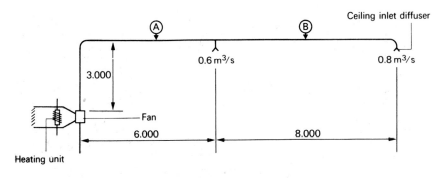

Fig. 10.9

12. State the fan laws for: (a) constant air density; (b) variable air density.

13. What information is required to be given to a manufacturer when ordering a fan for a certain system of ventilation?

14. A fan absorbs 1.25 kW of power and discharges 2.3 m³/s when the impeller angular velocity is 1300 revolutions per minute. If the impeller is reduced to 1000 revolutions per minute, calculate the discharge in m³/s and the power absorbed for this new condition.

Answers: 1.769 m³/s; 0.569 kW

15. A fan absorbs 1.5 kW when discharging 3 m³/s of air and operating at a total pressure of 400 Pa. Calculate the percentage efficiency of the fan.

Answer: 80 per cent

16. A fan develops a total pressure of 500 Pa when discharging air at a density of 1.2 kg/m³. If the temperature of the air is raised so that the air density becomes 0.95 kg/m³, calculate the pressure to be developed by the fan in order to discharge the same volume of air.

Answer: 395.833 Pa

Chapter 11

Air conditioning, principles and systems

Principles

The term 'air conditioning' has often been misused; for example, denoting a system of heating combined with mechanical ventilation. It should, however, be defined as a system giving automatic control, within predetermined limits of the environmental conditions, by heating, cooling, humidification, dehumidification, cleaning and movement of air in buildings. The control of these conditions may be desirable to maintain the health and comfort of the occupants, or to meet the requirements of industrial processes irrespective of the external climatic conditions.

Advantages

1. In factories and offices the working efficiency of personnel is improved and work output is known to increase. There is also a reduction in illness and absenteeism.
2. Shops and departmental stores have increased sales due to customers and staff being able to enjoy greater comfort. The cost of the plant will therefore be offset by extra income from customers.
3. There is a reduction in cleaning and decorating; also fabrics and furnishings last longer.
4. Hotels, restaurants, theatres and cinemas received better patronage.
5. Many industrial premises need air conditioning to keep plant and processes working at maximum efficiency; these include computer rooms, food production rooms, electronic laboratories, textile factories, printing rooms and laboratories.
6. Sealed windows reduce the entry of noise from aircraft and traffic; also the entry of fumes, smoke, dust and draughts.
7. There is less risk of fire due to static electricity caused by dry air.

Relative humidity

The relative humidity of the air is usually expressed as a per cent and is a ratio between the actual amount of moisture in a given volume of air, and the amount of moisture that would be necessary to saturate that volume. It may also be expressed as a ratio between the actual vapour pressure and the saturated vapour pressure. Most people feel comfortable when the relative humidity is between 30 and 70 per cent and an air conditioning plant usually operates to maintain a relative humidity of between 40 and 50 per cent. When the air is too dry, moisture evaporates more readily from the skin and this produces a feeling of chilliness, even if the air temperature is satisfactory. Dry air also removes moisture from the nose, throat and eyes, causing these to be irritated. When the air is too damp, moisture cannot readily evaporate from the skin and this causes the body to become overheated, resulting in a feeling of drowsiness.

$$\text{Relative humidity} = \frac{\text{mass of water vapour in a given volume of air}}{\text{mass of water vapour required to saturate the same volume of air at the same temperature}}$$

$$\text{Relative humidity} = \frac{\text{actual vapour pressure}}{\text{saturation vapour pressure at the original air temperature}}$$

The psychrometric chart

This shows, by graphical representation, the relationship that exists between wet and dry bulb temperatures at different relative humidities. The usefulness of the chart is extended by the addition of moisture content values and specific enthalpy, or total heat of air. The chart is based on a barometric pressure of 101.325 kPa. The following illustrates the use of the chart (see Fig.11.1).

1. *Dry bulb temperature:* read vertically downwards to 25 °C.
2. *Wet bulb temperature:* trace parallel to the sloping wet bulb lines to intersect the wet bulb temperature on the 100 per cent relative humidity curve at 18 °C.
3. *Relative humidity:* read directly from the point of intersection of the dry and wet bulb temperatures, the relative humidity curve at 50 per cent.
4. *Dew point:* travel horizontally to the left of the intersection, to intersect the 100 per cent curve at 14 °C.
5. *Moisture content:* read horizontally to the right at 0.010 kg/kg (dry air).
6. *Specific enthalpy or total heat:* this is found by drawing a line to the total heat lines at the chart extremities, which reads 50.1 kJ/kg.

Note: The total heat lines are not parallel to the wet bulb lines.

The following examples will show how the chart is used to solve various problems.

123

Example 1. *In winter, air at a dry bulb temperature of 5 °C and 60 per cent relative humidity enters a building through a heating battery and is heated to a dry bulb temperature of 20 °C without adding moisture.*

From the chart find:

1. Wet bulb temperature of the incoming air.
2. The relative humidity of the heated air.

Figure 11.2 shows how these values are found from the chart and it will be seen that the wet bulb temperature of the incoming air will be 2.2 °C and the relative humidity of the heated air 25 per cent.

Note: This is sensible heating of the air and it shows that without humidification the relative humidity of the air is too low.

Example 2. *In summer, air at a dry bulb temperature of 25 °C and a wet bulb temperature of 21 °C enters a building through a cooling coil and is cooled to a dry bulb temperature of 20 °C.*

From the chart find:

1. The relative humidity of the incoming air.
2. The relative humidity of the supply air after cooling.

Figure 11.3 shows how these values are found from the chart and it will be seen that the relative humidity of the incoming air is 70 per cent and the relative humidity of the supply air after cooling is 95 per cent.

Note: This is sensible cooling of the air in summer and it shows that without dehumidification the relative humidity of the supply air is too high.

Need for humidifying and dehumidifying

It will be clear from Figs 11.2 and 11.3 that if air enters a building at a low temperature in winter and is passed through a heating battery, its relative humidity will be reduced and may be below that required for human comfort. Also if air enters a building at a high temperature in summer and is passed through a cooling battery and cooled above its dew point, its relative humidity may be increased above that required for human comfort.

Example 3. *The air in a room has a dry bulb temperature of 22 °C and a wet bulb temperature of 16 °C.*

From the chart find:

1. The relative humidity of the air.
2. The temperature of the walls when condensation would occur.

Figure 11.4 shows how these values are found from the chart and it will be seen that the relative humidity is 52 per cent and condensation will occur when the temperature of the walls reach dew point at 12.5 °C.

124

Dehumidification by cooling and reheating

In summer, the air passing through an air conditioning plant is cooled in the spray to below its dew point and then reheated.

Example 4 (see Fig. 11.5). *Air enters the plant at a dry bulb temperature of 25 °C and 70 per cent relative humidity and is required to be cooled to a dry bulb temperature of 20 °C and 50 per cent relative humidity.*

From the chart find (assuming washer efficiency of 100 per cent):

1. The temperature of air in the washer.
2. The reduction in the moisture content of the supply air.

Note: The air is first cooled to below a wet bulb temperature of 10 °C in the spray and then reheated to 20 °C when since no more moisture is added, the relative humidity will have been reduced to 50 per cent.

Example 5 (see Fig. 11.6). *If air enters the plant in winter at a dry bulb temperature of 10 °C and 60 per cent relative humidity, the amount of moisture in the air would be very low and if the air was heated (without adding moisture) to a supply temperature of 25 °C dry bulb, its relative humidity would be reduced to about 22 per cent. If the air is preheated to a dry bulb temperature of 20 °C before entering the washer, the resultant condition would be a wet bulb temperature of 11.2 °C, and if the air is reheated to a dry bulb temperature of 25 °C its relative humidity would be about 41 per cent, which has increased the relative humidity from 22 to 41 per cent.*

Adiabatic saturation

Evaporation of moisture in the air takes place without a change of temperature and if unsaturated air is passed over a thin film of water the air will evaporate moisture for as long as it remains in an unsaturated state. If there is no other source of heat, the latent heat of evaporation will be supplied from the air and its temperature will be lowered.

In Example 5 air at a dry bulb temperature of 10 °C and 60 per cent relative humidity is passed through a washer with 100 per cent efficiency and not preheated; its resultant condition would be a wet bulb temperature of 6.5 °C. This process is known as 'adiabatic saturation', being a term used to describe a process where no heat is added or taken from the air.

Air conditioning systems

The type of system depends upon the type of building, and if it is necessary to vary the air temperatures and relative humidities also upon the space within the building for the plant ductwork and pipework.

The systems may be divided into three categories:

1. All air systems where the conditioned air is treated in a central plant and ducted to the various rooms. It requires large duct spaces and plant rooms, but very little is taken up inside the rooms.

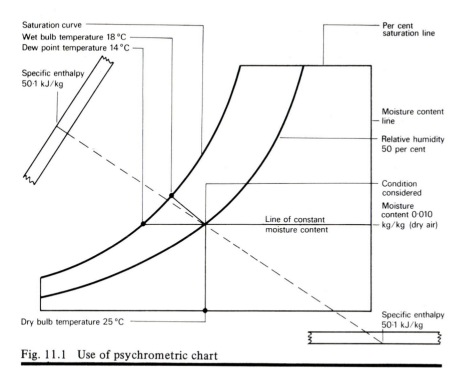

Fig. 11.1 Use of psychrometric chart

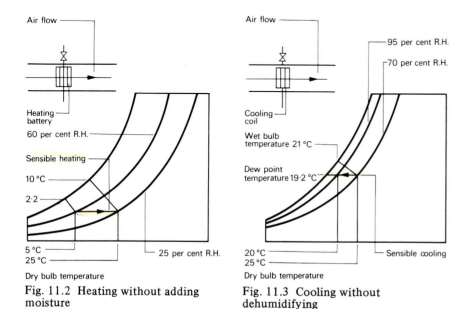

Fig. 11.2 Heating without adding moisture

Fig. 11.3 Cooling without dehumidifying

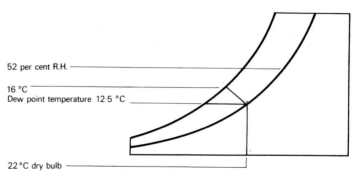

Fig. 11.4 Condensation on wall surface

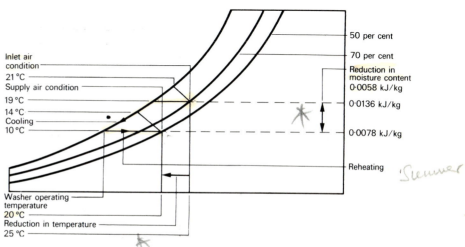

Fig. 11.5 Dehumidification — cooling, washing and reheating

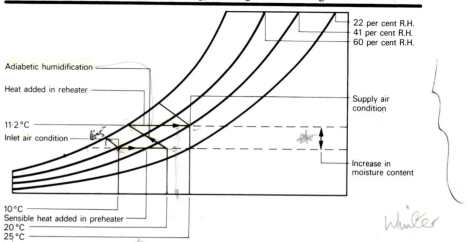

Fig. 11.6 Humidification — preheating, washing and final heating

2. Air and water systems in which the air is treated in a central plant, but the bulk of heating and cooling is done in the rooms by passing the air over hot or cold water coils supplied from a central boiler and refrigeration plant. The central plant is smaller and less ductwork is required, but space will be taken up by the room air heating and cooling units.

3. Self-contained units in which the complete air conditioning process is carried out and sited inside the various rooms against the outside wall, so that fresh air can be drawn into the units through a short duct in the wall. There is less duct space and the central plant is small or non-existent, but some space is taken up inside the room for the units.

Central plant system

This is suitable for large spaces where the air temperature and relative humidity is constant. The system is suitable for large factory spaces, open plan offices, theatres, cinemas, supermarkets and assembly halls. The main air conditioning unit may be sited at ground floor level, in a basement, or in a plant room on the roof.

Figure 11.7 shows a longitudinal diagram of a central plant system, which operates as follows:

1. Fresh air is drawn in from the side of the building where the air is likely to be the cleanest and at this stage is mixed with a proportion of recirculated air, which will reduce the load on the heating or cooling plants.

2. The air passes through a dry or viscous filter to remove suspended matter in the form of dust. If required, the air can be further cleaned by passing it through an electrostatic filter.

3. In winter, the air is heated by a preheating coil formed of finned copper pipe, heated by steam, electricity or hot water. Preheating of the air will allow the air to absorb more moisture in the washer and also prevent freezing of the water in the sprays, should the temperature of the incoming air fall below 0 °C. The spray water itself may also be heated.

4. In summer, the air is cooled by a cooling coil, or by passing the air through cold water in the spray. The cooling coil or spray water may be cooled by a refrigeration plant or by water pumped from a deep borehole. The air at this stage is cooled to below dew point and will therefore be saturated. The washer also cleans the air. The warming or cooling of the water in the washer forms an important factor in the air conditioning process. Provision is made for changing the spray water and cleaning out the cistern when necessary.

5. After leaving the washer, the air passes through two sets of glass or galvanised steel plates, made up in a zig-zag formation. The first set is known as scrubber plates which are washed by a continuous stream of water, so that any particles of dust still held in suspension are washed down to the cistern. The second set are known as eliminator plates and are designed to intercept any droplets of water held in the air, so that only absorbed moisture is carried forward with the air to the final or reheater.

6. In both winter and summer the air may require reheating in the final heater, which brings the air up to the required temperature with a corresponding reduction in the relative humidity. This heater is similar in construction to the preheater, but has a greater surface area.

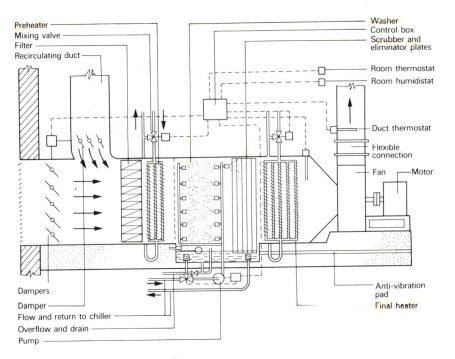

Fig. 11.7 Air conditioning central plant unit

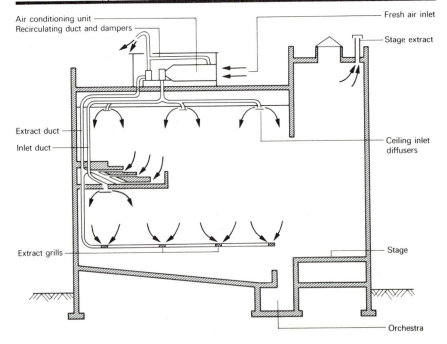

Fig. 11.8 A schematic diagram of an air conditioning plant for a theatre

7. The inlet fan forces the conditioned air into the building via the inlet ductwork and diffusers.

8. Vitiated air is extracted from the building by the extract fan via the extract ductwork and grills. The extract fan is smaller than the inlet fan and this causes slight pressurisation of the air inside the building, which helps to prevent the entry of draughts and dust.

If required, up to about 75 per cent of the extract air may be recirculated via the recirculating duct, also the air may be drawn through the lighting fitting, which would extract heat from the lamps and also improve lighting efficiency by about 14 per cent.

Automatic control

Motorised dampers automatically control the flow of air through the fresh air inlet and recirculating duct. The temperatures of the preheater, washer and reheater are thermostatically controlled, to give the required supply air temperature and relative humidity.

Figure 11.8 shows a central plant system installed in a theatre, using downward air distribution.

Variable air volume system

In this system, the air is supplied from a central plant unit at a temperature and relative humidity, which vary with the weather conditions. An insulated single duct carries the conditioned air at high velocity to variable volume units, usually positioned in the ceiling. The units are provided with thermostatically controlled actuators, which vary the quantity of air supplied to each space. The amount of air passing into the room from the units, vary with the room temperature and the system is particularly suitable for a building having fairly evenly distributed cooling loads.

The dual duct system of air conditioning

This system is similar to the central plant system, in that it uses air as a heating or cooling medium and there is no pipework in the rooms being air conditioned. The system has two ducts, for hot and for cold air respectively. The use of these air ducts provides a means of controlling varying temperatures in different rooms and it is possible to meet the most exacting room climatic requirements.

All the main air conditioning units can be housed in the plant rooms and the hot and cold air ducts may be installed in the ceiling void to mixing or blending units. The air in the ducts flows at a higher pressure and velocity than the central plant system and this allows the ducts to be smaller. They must be fully airtight, or otherwise there is a risk of whistling noises resulting from small leaks. Circular ducts are easier to seal than rectangular ones, but because of the greater size they are more difficult to house in the ceiling void.

Because of the higher pressure and velocity of the air, the fan must operate at a higher speed and this requires a larger fan motor, which uses more electrical power. The system requires more duct space than the central plant system, but this can usually be arranged in a false ceiling, which will take up to 300 mm from the head room of each storey. If required, the units may be installed under

the windows, to intercept the cold air in winter, or inside the ceiling void, which supplies air at the desired temperature to branch ducts and ceiling diffusers. Alternatively, the ceiling void itself may be used for the supply air and the air forced through perforated ceiling panels. The ducts must be thermally insulated, to prevent heat loss from the hot air duct and heat gain to the cold air duct.

Besides having the advantage of providing variations of temperatures, the system also provides varying ventilation rates and rooms requiring different air changes can be served from the same plant. The system, however, does not operate on a constant relative humidity principle, and this has to be sacrificed for the advantage of varying temperatures and ventilation rates.

Operation of the plant

Fresh air from the cleanest side of the building is drawn into a main treatment plant and is mixed as required with recirculating air. Dust is removed in the filter and the air is preheated and humidified in winter. In summer, the air is filtered, washed and cooled in the cooling coil. In both winter and summer, the air is forced by a fan into two ducts; a hot air duct is fitted with a heating battery, and a cold-air duct fitted with a cooling coil or chiller. After passing through the heater and or chiller, the air passes via a cold air and a hot air duct to mixing or blending units, usually installed under the windows of the various rooms.

The mixing of the hot or cold air streams passing through the mixing units is controlled by a room thermostat, so that the supply air to the rooms may be set and maintained at any temperature within the limits of the hot or cold air streams. The unit also incorporates a means of automatically controlling the volume of air passing through, regardless of the variations of air pressure in the ducts. This is known as a 'constant volume controller' and this is an essential part of the dual duct system.

Figure 11.9 shows a schematic diagram of a dual duct system, where the vitiated air is extracted through the light fittings.

Figure 11.10 shows a sectional elevation of one type of constant volume room mixing or blending unit, which is fitted below a window.

Figure 11.11 shows an elevation of the mixing unit, with the method of making the flexible connections from the main hot and cold air ducts.

Figure 11.12 shows how mixing units may be installed, to force the conditioned air through a perforated ceiling into a room.

The induction convector air conditioning system

Most rooms in a building have different requirements so that the supply air will have to be conditioned to meet the comfort needs of each room. The induction system can be used as an alternative to the dual duct system for varying the supply air temperatures in multi-roomed buildings. Instead of hot and cold air ducts required for the dual duct system, the induction system utilises hot and or cold water pipes, which takes up less space, but a single duct is also required for the primary air from a main conditioning unit.

An induction system, as the name implies, causes the secondary air in the room to be recirculated by the primary air passing through nozzles in an induction unit, and this causes good air movement in the room. This induced air flows over a cooling or heating coil and cooled or heated secondary room air is

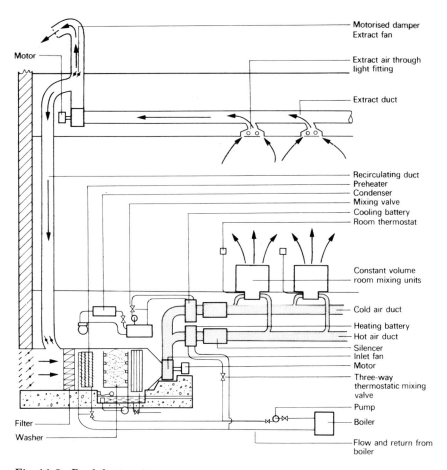

Fig. 11.9 Dual duct system

Labels (Fig. 11.9):
Motorised damper
Extract fan
Motor
Extract air through light fitting
Extract duct
Recirculating duct
Preheater
Condenser
Mixing valve
Cooling battery
Room thermostat
Constant volume room mixing units
Cold air duct
Heating battery
Hot air duct
Silencer
Inlet fan
Motor
Three-way thermostatic mixing valve
Pump
Boiler
Flow and return from boiler
Filter
Washer

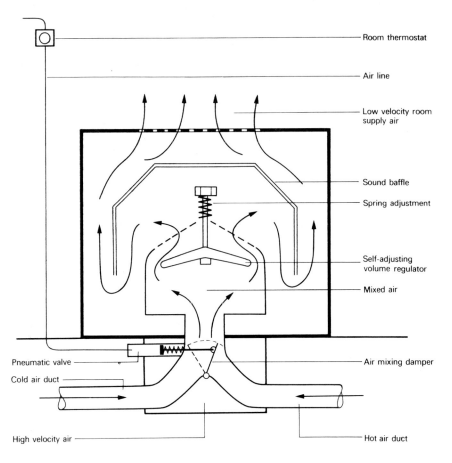

Fig. 11.10 Constant volume mixing unit

Labels (Fig. 11.10):
Room thermostat
Air line
Low velocity room supply air
Sound baffle
Spring adjustment
Self-adjusting volume regulator
Mixed air
Air mixing damper
Pneumatic valve
Cold air duct
High velocity air
Hot air duct

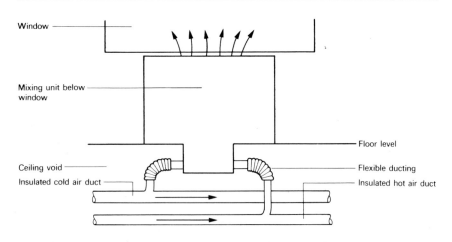

Fig. 11.11 Elevation of the mixing unit

Labels (Fig. 11.11):
Window
Mixing unit below window
Ceiling void
Insulated cold air duct
Floor level
Flexible ducting
Insulated hot air duct

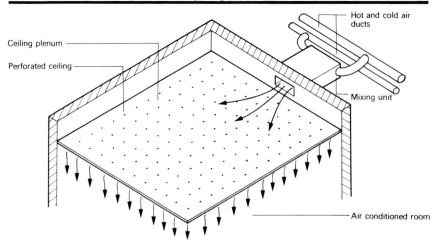

Fig. 11.12 Installation of the mixing unit

Labels (Fig. 11.12):
Hot and cold air ducts
Ceiling plenum
Perforated ceiling
Mixing unit
Air conditioned room

mixed with the primary conditioned air before being discharged through a grill at the top of the induction unit.

A two-pipe induction system is the most common type of air/water system in use. This system operates on the two-pipe principle, and is referred to as 'the two pipe change over induction convector system'. Flow and return pipes are used leading to induction unit convectors, one inlet and one return pipe. During summer, chilled water is pumped from a refrigerating unit, whilst during winter hot water is pumped from a boiler plant. The system can easily be changed over from heating to cooling and vice versa, depending upon the external climatic conditions. The change over can be effected manually, or automatically from a control station.

With the aid of the system the rooms can be heated or cooled quite apart from the limited amount of heating or cooling by the primary air from the main air conditioning unit. With rooms of differing aspects, the water piping may be so arranged that if required chilled water may be pumped through to rooms having a southern aspect, and heated water pumped through to rooms having a northern aspect.

During the spring and autumn transitional periods, an important factor with the system is the date when a change over is made, from winter or summer operation and vice versa, and this decision would be left to the building owner. In theory the change over from winter to summer operation can be made when the heat gains from the sun, and heat generated from inside the building from people, machinery and lighting is equal to the heat losses through the structure to the outside atmosphere.

Operation of the plant

The air is treated in the main air conditioning unit as previously described. The inlet fan then forces air through the primary air duct to the various induction units fitted below the windows. In winter, the curtain of cold air descending past the windows can in this way be drawn into the induction unit and a warm air curtain formed over the window opening, so that the window area is pleasantly warmed. In summer, a stream of cold air blown vertically upwards cools the area around the window opening and reduces the solar heat gains inside the room.

In order to keep the duct size down to a minimum and to create sufficient air velocity in the induction unit, the air velocity in the duct is in the order of 20 m/s, whereas with the ordinary air conditioning central plant system this velocity is about 5 m/s. In summer, chilled water is circulated through finned copper pipe coils, fitted inside the induction units; during winter, hot water is circulated through these coils. The primary air, at high velocity, passes through induction nozzles inside the induction unit and in so doing reduces the air pressure below these nozzles. The secondary air inside the room is thus drawn into the unit and is mixed with the primary air before being discharged into the room.

If the occupants of the room are satisfied with the temperature of the primary air, a damper can be arranged so that the cold or hot coil is bypassed and only primary air is then used. If it is required to warm or cool the primary air, depending upon the time of year and use of room, the damper can be arranged so that secondary air from the room passes through the cold or hot coil and the cooled or heated secondary air will mix with the primary air. The mixture of secondary to primary air in the induction unit is usually three to six

volumes of secondary air, to one volume of primary air. Air in the room can be circulated up to a maximum of 6 m from the induction units; rooms having a greater depth than 6 m will require a supplementary air supply. Vitiated air in the room is extracted at some convenient point, usually at ceiling level, and in order to reduce the load in winter the air can be extracted through the electric light fitting. In some induction systems extracted air is not recirculated back to the main unit, but is discharged to the atmosphere. Increasing costs of fuel and energy makes this method uneconomic and leads to higher heating and cooling loads. Other types of more complex induction systems include:

1. The three-pipe induction system, in which both cooling water and heating water are fed into the induction unit and a common return pipe returns both the heated and cooled water back to the main plant.
2. The four-pipe induction system, in which both cooling water and heating water are fed into the induction units and separate return pipes, returns water back to the main plant. In this way the regulation instability and mixing losses experienced with the three-pipe system are obviated.

Both these systems permit each induction unit to give full heating and cooling all the year round and there is no need to change over from heating or cooling or vice versa, as with the two-pipe change over system. These systems, however, are more costly to install and would only be used for buildings such as first-class hotels.

Figure 11.13 shows the installation of the system for a two-storey building having separate zones. For clarity, only the primary air duct and the main air conditioning unit is shown.

Figure 11.14 shows the same system connected with pipework from a heater and a cooling battery.

Figure 11.15 shows a section of an induction unit and the principle of operation.

Figure 11.16 shows the installation of an induction unit in a room.

Self-contained air conditioners

There are usually for smaller buildings such as houses, small shops, offices, restaurants, clubs and hotels. They are a complete unit and only require electrical and perhaps water connections. Sizes vary from small units for one room to larger sizes for several rooms served from ductwork.

The smaller units contain a heater and a refrigerator, or a heat pump with compressor, fan or fans, evaporator or cooling coil, condenser or heating coil, filter and, if required, a humidifier. They may be mounted against the outside wall so as to draw in fresh air through a short duct at the rear of the unit.

The larger units contain heating coils, which may be heated directly by electricity or indirectly by hot water, cooling coils from a refrigerator or cooling tower, filter, fan and again, if required, means of humidification. These units usually require ducting for the distribution of the conditioned air.

Several manufacturers produce modules of various parts of the unit and it is therefore possible to install a unit with only a heating coil and filter; a cooling coil and humidifier may be added later. These modules are of standard sizes and fit into the unit by sliding or bolting into place and connecting the necessary

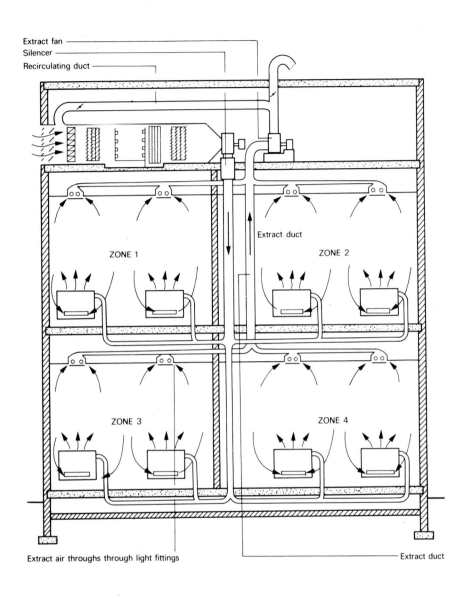

Extract fan
Silencer
Recirculating duct

Extract duct

ZONE 1

ZONE 2

ZONE 3

ZONE 4

Extract air throughs through light fittings

Extract duct

Fig. 11.13 Induction convector – air conditioning showing induction units and primary air supply

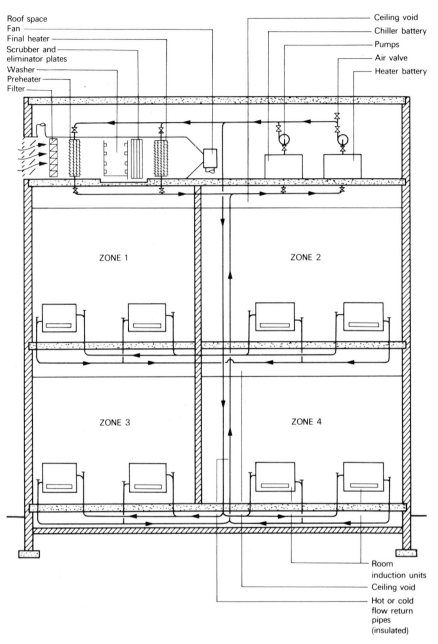

Roof space
Fan
Final heater
Scrubber and eliminator plates
Washer
Preheater
Filter

Ceiling void
Chiller battery
Pumps
Air valve
Heater battery

ZONE 1

ZONE 2

ZONE 3

ZONE 4

Room induction units
Ceiling void
Hot or cold flow return pipes (insulated)

Fig. 11.14 Induction convector – air conditioning showing induction units and water supply pipes

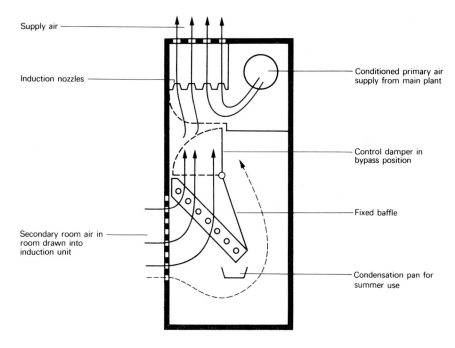

Supply air

Induction nozzles

Conditioned primary air supply from main plant

Control damper in bypass position

Fixed baffle

Secondary room air in room drawn into induction unit

Condensation pan for summer use

Fig. 11.15 Vertical section through an induction unit

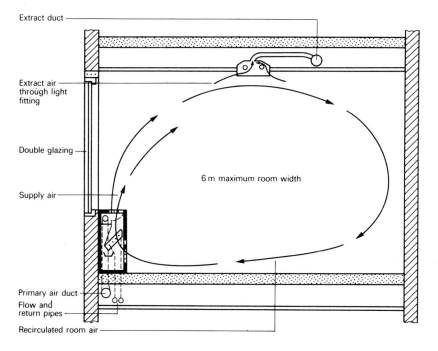

Extract duct

Extract air through light fitting

Double glazing

Supply air

6 m maximum room width

Primary air duct
Flow and return pipes

Recirculated room air

Fig. 11.16 An induction unit fitted in a room

pipework or cables. The heat pump has the advantage of providing both heating and cooling from the same system and it is cheaper to run than separate heating and cooling units. It may be used to operate both small and larger units.

The heat pump

The name is derived from the fact that the heat pump takes heat from a relatively cool body and raises it to a useful temperature, similar in a way to a mechanical pump, raising water to a higher level. It acts in the opposite way to a heat engine, in that it takes in heat at low temperature together with mechanical power and gives out heat at high temperature, whereas the heat engine takes in heat at high temperature and gives out mechanical power and heat at low temperature. It utilises the normal refrigeration cycle to absorb heat from one place and release it for use in another, and can therefore be used to reduce the temperature by forcing cold air into a building in summer, or to increase the temperature by forcing warm air into a building in winter.

Operation

As a vapour, a refrigerant has latent heat and absorbs latent heat as it condenses the temperature at which the change of state occurs, depending upon the pressure. At low pressure, the change takes place at low temperature and at high pressure at high temperature. Thus a liquid can be made to boil at a very low temperature, or a vapour can be made to condense at a high temperature, merely by varying the pressure.

The refrigeration cycle

The refrigerant is compressed in the compressor and is liquefied in the condenser; this liquid gives off latent heat which is extracted from the condenser coils by a fan, natural convection or water. The liquid then passes through an expansion or pressure-reducing valve, where it is reduced in pressure thus allowing the liquid to vaporise. This vaporisation requires latent heat from the air surrounding the evaporator coils and this causes the air to be cooled. The cycle is shown in Fig. 11.17.

Heat pump cycle

The refrigerator extracts heat from the air in the evaporator and gives off heat in the condenser. In a refrigerator this heat is wasted to the air or water, but it can be used as a source of heat for heating of air or water. When the condenser is used for heating purposes, the system is known as a heat pump. If the system is used in a building, the evaporator can be used for cooling in summer and the condenser for heating in winter. The cycle is shown in Fig.11.18.

Coefficient of performance (COP)

The coefficient of performance can be expressed as follows:

$$COP = \frac{tc}{tc\ te}$$

where te = Evaporator temperature in degrees Kelvin

tc = Condenser temperature in degrees Kelvin

It will be clear from the above equation that, in common with all vapour

131

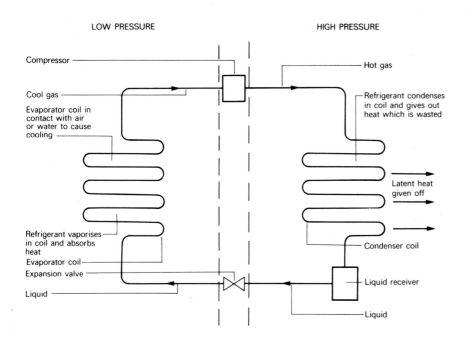

Fig. 11.17 Vapour refrigerator cycle

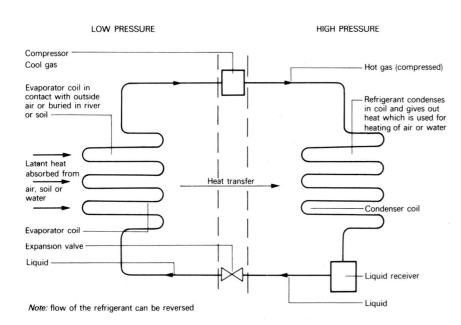

Fig. 11.18 Heat pump cycle

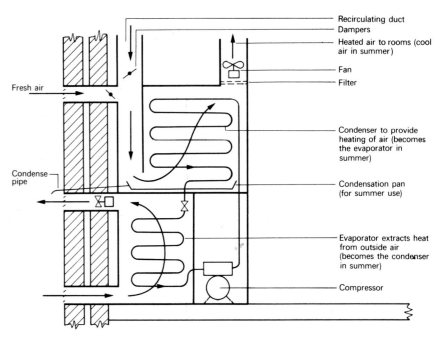

Fig. 11.19 Detail of heat pump for winter use (refrigerant flow reversed for summer use)

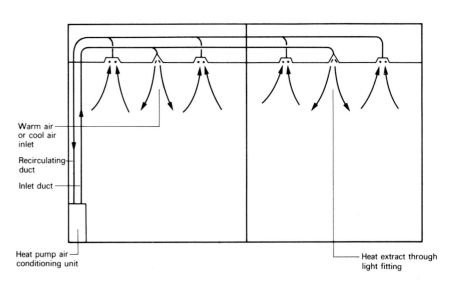

Fig. 11.20 Use of the heat pump (detail of inlet and recirculating ductwork)

refrigeration systems, the heat pump operates to a greater advantage when working with a low condenser temperature and a higher evaporator temperature.

Note: The coefficient of performance is not the efficiency of the heat pump.

Example. *Calculate the COP of a heat pump when the condenser and evaporator temperatures are 45 °C and 4 °C respectively, assuming 100 per cent efficiency.*

Condenser temperature = 45 °C + 273

$$tc = 318 °K$$

Evaporator temperature = 4 °C + 273

$$te = 277 °K$$

Then COP $= \dfrac{318}{318 - 277}$

COP $= 7.75$

This means that for every 1 kW of electrical power used there would be a heat output of 7.75 kW if the machine were 100 per cent efficient. The large heat output gives the impression that the heat pump is an energy-creating device, but the machine, however, only takes existing heat energy at low temperature from any available source and raises it to a higher temperature. Allowing for loss of efficiency, the machine would show a coefficient of performance of between two and three.

It may be wondered why, with such a high performance, the heat pump is not used more widely. There are two possible reasons for the apathy shown towards the heat pump:

1. Low outside air temperature in winter months.
2. Higher cost of plant compared with traditional systems.

With the increasing cost of fuels, however, the heat pump as a big advantage for air conditioning by providing heating in winter from the condenser and cooling in summer from the evaporator at competitive costs.

The low-grade heat for the evaporator may be obtained by the following methods:

1. By laying the coils at the bottom of a river or stream and extracting heat from the water.
2. By burying the coils below ground and extracting heat from the soil.
3. By drawing atmospheric air over the evaporator coil and extracting heat from the air. Figure 11.19 shows a unit air conditioner, utilising the heat pump cycle to warm a building in winter by extracting heat from the low temperature outside air. The same unit may be used for cooling rooms in summer by reversing the flow of refrigerant in the heat pump circuit. Figure 11.20 shows the method of installing the unit in a building.

To supplement the heat output from the condenser coils during very cold spells, a boost heater may be incorporated in the unit, which is switched on automatically when the outside air temperature falls below a required level. This supplementary heating will only be in use for a very small part of the heating season, depending upon the degree of thermal insulation.

Figure 11.21 shows how a heat pump-operated air conditioner is fitted below the window opening and a boost heater has been incorporated.

Figure 11.22 shows a large modular type of air conditioning plant that is manufactured in a large range of sizes which will be suitable for most types of buildings.

Refrigerants

Liquids which boil at low pressures and temperatures are used for refrigerants, and the following have all been used in the past: ammonia, carbon dioxide, sulphur dioxide and methyl chloride.

Halogenated hydrocarbons (Freons) are now nearly always used due to their many advantages over the previously used liquids. They are: non-inflammable, non-toxic, non-explosive, odourless and operate at moderate pressures.

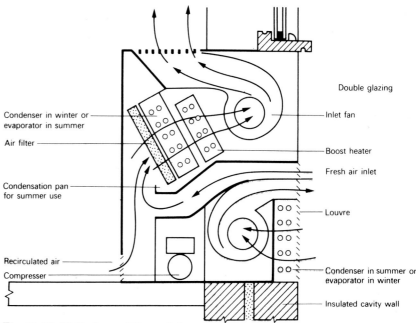

Fig. 11.21 Unit air conditioner below window

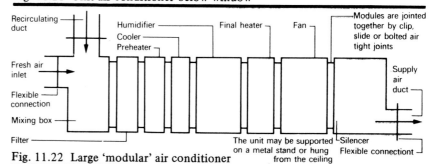

Fig. 11.22 Large 'modular' air conditioner

133

Appendix
SI Units

SI units (International System of units)

In 1971 the Council of Ministers of the European Economic Community (EEC) decided to commit all member countries to amend their legislation in terms of SI units. The United Kingdom had already decided that SI units would become the primary system of measurement, and legislation is established in some twenty-five countries, including Germany, France, India, the USSR and Czechoslovakia. The system is also being considered in the USA where a 3-year study has been completed on behalf of the Department of Commerce.

Base units

The SI system is based on seven units.

Quantity	Unit	Symbol
Length	metre	m
Mass	kilogram	kg
Time	second	s
Electric current	ampere	A
Thermodynamic temperature	kelvin	K
Luminous intensity	candela	cd
Amount of substance	mole	mol

Note: For ordinary temperature and the difference between two temperatures, i.e. temperature interval, the degree Celsius (°C) is used.

Supplementary units

Quantity	Unit	Symbol
Plane angle	radian	rad
Solid angle	steradian	sr

The radian is the angle between two radii of a circle which cut off on the circumference an arc equal in length to the radius. The steradian is an angle which having its vertex in the centre of a sphere, cuts off an area of the surface of the sphere equal to that of a square having sides of length equal to the radius of the sphere.

Derived units

These are expressed algebraically in terms of base units and for supplementary units

Quantity	Name of derived units	Symbol	Units involved
Frequency	hertz	Hz	$1 \text{ Hz} = 1 \text{ s}^{-1}$ (1 cycle per second)
Force	newton	N	$1 \text{ N} = 1 \text{ kg m/s}^2$
Pressure and stress	pascal	Pa	$1 \text{ Pa} = 1 \text{ N/m}^2$
Work, energy, quantity of heat	joule	J	$1 \text{ J} = 1 \text{ N m}$
Power	watt	W	$1 \text{ W} = 1 \text{ J/s}$
Quantity of electricity	coulomb	C	$1 \text{ C} = 1 \text{ A s}$
Electric potential, potential difference, electromotive force	volt	V	$1 \text{ V} = 1 \text{ W/A}$
Electric capacitance	farad	F	$1 \text{ F} = 1 \text{ As/V}$
Electric resistance	ohm	Ω	$1 \text{ }\Omega = 1 \text{ V/A}$
Electric conductance	siemens	S	$1 \text{ S} = 1 \text{ }\Omega^{-1}$
Magnetic flux, flux of magnetic induction	weber	Wb	$1 \text{ Wb} = 1 \text{ V s}$
Magnetic flux density, magnetic induction	tesla	T	$1 \text{ T} = 1 \text{ Wb/m}^2$
Inductance	henry	H	$1 \text{ H} = 1 \text{ Vs/A}$
Luminous flux	lumen	lm	$1 \text{ lm} = 1 \text{ cd sr}$
Illuminance	lux	lx	$1 \text{ lx} = 1 \text{ lm/m}^2$

Multiples and sub-multiples of SI units

Factor		Prefix Name	Symbol
One billion	10^{12}	tera	T
One thousand million	10^9	giga	G
One million	10^6	mega	M
One thousand	10^3	kilo	k
One hundred	10^2	hecto	h
Ten	10	deca	da
One-tenth	10^{-1}	deci	d
One-hundredth	10^{-2}	centi	c
One-thousandth	10^{-3}	milli	m
One millionth	10^{-6}	micro	μ
One thousand millionth	10^{-9}	nano	n
One million millionth	10^{-12}	pico	p

Units for general use

Quantity	Unit	Symbol	Definition
Time	minute	min	1 min = 60 s
	hour	h	1 h = 60 min
	day	d	1 d = 24 h
Plane angle	degree	°	$1° = (\pi/180)$ rad
	minute	′	$1′ = (1/60)°$
	second	″	$1″ = (1/60)′$
Volume	litre	l	$1\,l = 1\ dm^2$
Mass	tonne	t	$1\ t = 10^3\ kg$

Bibliography

Relevant BS. British Standards Institution.

Relevant BSCP. British Standards Institution.

Building Regulations 1976. HMSO.

Relevant BRE Digests. HMSO.

Electrical Regulations. 14th edition. The Institution of Electrical Engineers.

Fire Protection Handbook. Mather & Platt Ltd.

Electrical Services in Buildings. The Electricity Council.

Electrical Installations in Buildings. J. A. Crabtree & Co. Ltd.

D. C. Pritchard. *Lighting.* Longman Group Ltd.

Smoke Control in Covered Shopping Malls. The Fire Research Association.

Lifts. Hammond and Clampness.

IHVE Guide Book A. The Institution of Heating and Ventilating Engineers.

The Architects Specification. The Architectural Press.

O. Faber and J. R. Kell. *Heating and Air Conditioning of Buildings.* The Architectural Press.

Steam Heating. Spirax Sarco Limited.

R. Chudley. *Construction Technology Volume 3.* Longman Group Limited.

P. Burberry. *Environment and Services.* Batsford Limited.

C. R. Bassett and M. Pritchard. *Heating.* Longman Group Limited.

Relevant manufacturers' catalogues contained in the Barbour Index and Building Products Index Libraries.

Index